METHODS AND PROTOCOLS IN FOOD SCIENCE

Series Editor
Anderson S. Sant'Ana
University of Campinas, Campinas,
Brazil

For further volumes:
http://www.springer.com/series/16556

Methods and Protocols in Food Science series is devoted to the publication of research protocols and methodologies in all fields of food science.

Volumes and chapters will be organized by field and presented in such way that the readers will be able to reproduce the experiments in a step-by-step style. Each protocol will be characterized by a brief introductory section, followed by a short aims section, in which the precise purpose of the protocol will be clarified.

Citrus Fruit Processing By-Products

Technologies and Basic Protocols

Edited by

Shweta Suri

School of Food Science, Washington State University, Pullman, USA; School of Health Science and Technology, University of Petroleum and Energy Studies (UPES), Dehradun, Uttarakhand, India

Anupama Singh

Food Engineering, National Institute of Food Technology Entrepreneurship and Management (NIFTEM), Sonipat, Haryana, India

Kandi Sridhar

INRAe, L'Institut Agro Rennes Angers, Rennes, France

Humana Press

Editors
Shweta Suri
School of Food Science
Washington State University
Pullman, USA

School of Health Science
and Technology
University of Petroleum
and Energy Studies (UPES)
Dehradun, Uttarakhand, India

Anupama Singh
Food Engineering
National Institute of Food Technology
Entrepreneurship and Management (NIFTEM)
Sonipat, Haryana, India

Kandi Sridhar
INRAe, L'Institut Agro Rennes Angers
Rennes, France

ISSN 2662-950X ISSN 2662-9518 (electronic)
Methods and Protocols in Food Science
ISBN 978-1-0716-5087-5 ISBN 978-1-0716-5088-2 (eBook)
https://doi.org/10.1007/978-1-0716-5088-2

This Humana imprint is published by the registered company Springer Science+Business Media, LLC, part of Springer Nature.
The registered company address is: 1 New York Plaza, New York, NY 10004, U.S.A.

Preface to the Series

The Methods and Protocols in Food Science series is devoted to the publication of research protocols and methodologies in all fields of food science. The series is unique as it includes protocols developed, validated, and used by food and related scientists as well as provides theoretical basis for each protocol. Aspects related to improvements in the protocols and adaptations and further developments in the protocols may also be approached.

The Methods and Protocols in Food Science series aims to bring the most recent developments in research protocols in the field as well as very well-established methods. As such the series targets undergraduates, graduates, and researchers in the field of food science and correlated areas. The protocols documented in the series will be highly useful for scientific inquiries in the field of food sciences and presented in such a way that the readers will be able to reproduce the experiments in a step-by-step style.

Each protocol will be characterized by a brief introductory section, followed by a short aims section, in which the precise purpose of the protocol is clarified. Then, an in-depth list of materials and reagents required for employing the protocol is presented, followed by comprehensive and step-by-step procedures on how to perform that experiment. The next section brings the dos and don'ts when carrying out the protocol, followed by the main pitfalls faced and how to troubleshoot them. Finally, template results will be presented and their meaning/conclusions addressed.

The Methods and Protocols in Food Science series will fill an important gap, addressing a common complain of food scientists, regarding the difficulties in repeating experiments detailed in scientific papers. With this, the series has a potential to become a reference material in food science laboratories of research centers and universities throughout the world.

Campinas, Brazil ***Anderson S. Sant'Ana***

Preface

Citrus fruit by-products, namely, peels (flavedo and albedo), seeds, and pomace, are regarded as valuable bio-resource materials with multifaceted applications. These by-products are abundant in bioactive compounds that offer significant health benefits and are ideal for developing specialized food products. Given the significant proportion of these by-products and their rich nutrient profiles, comprehensive knowledge of the analytical methods for assessing citrus peels, seeds, and pomace is crucial for food science professionals. Hence, this handbook aims to furnish researchers with advanced methodologies and procedures pertinent to the analysis of citrus by-products. Each chapter offers comprehensive insights into the techniques and protocols for the extraction of bioactive compounds, as well as the determination of phenols, flavonoids, carotenoids, dietary fiber, pectin, organic acids, and essential oils present in citrus by-products. The handbook extensively covers reagent and sample preparation, analytical instruments employed in conducting research, and safety considerations. Its primary objective is to facilitate the strategic utilization of citrus by-products obtained from the citrus fruit-processing sector. The main goal is to create a pathway for improved utilization of citrus by-products in alignment with the United Nations Sustainable Development Goal No. 12: Responsible production and consumption.

Chapter 1 provides an overview of citrus by-products, discussing their market potential and applications. It emphasizes the effective use of these by-products to reduce waste and promote sustainable practices, which is important for food security. Chapter 2 focuses on methods to extract bioactive compounds from citrus by-products. It covers both traditional and new green extraction methods, along with their advantages and disadvantages. Chapter 3 explains procedures to determine total phenols in citrus by-products. It describes standard methodology for estimating total polyphenol content, including its benefits and limitations. Chapter 4 discusses methods to determine total flavonoids in citrus by-products. It includes various extraction techniques, like solvent extraction, followed by quantification using colorimetric, HPLC, or spectrophotometric methods. Chapter 5 explores the methods for estimating antioxidants in citrus by-products. It presents various in vitro assays, including DPPH, FRAP, ABTS, TEAC, and ORAC. Chapter 6 discusses procedures to estimate dietary fibers in citrus by-products, focusing on an enzymatic-gravimetric method to measure total dietary fiber. Chapter 7 deals with methods to quantify carotenoids in citrus by-products. Chapter 8 covers approaches for measuring pectin content in citrus peels. It discusses advanced techniques like FTIR and HPLC that provide valuable knowledge of the pectin's structure and function. Chapter 9 explores methods for separating and identifying organic acids in citrus by-products. It discusses reliable techniques such as high-performance liquid chromatography with photo-diode array detection, gas chromatography with mass spectrometry, and ion chromatography. Chapter 10 describes a protocol for extracting, separating, and quantifying naringin in citrus by-products using HPLC. This chapter presents a quick and simple method for reverse phase separation and measurement of naringin using HPLC-UV, assisted by ultrasonic extraction. Chapter 11 explores the procedures to analyze functional groups in citrus by-products using Fourier transform infrared spectroscopy. Chapter 12 discusses methods for identifying essential oils in citrus seeds. It highlights extraction methods, including solvent-based techniques, and advanced analysis tools like gas chromatography, gas chromatography-mass spectrometry, high-performance liquid chromatography, and enzyme-linked immunosorbent assay.

This book serves as a valuable resource for food scientists, researchers, academicians, principal investigators, horticulturists, agriculturists, pharmacologists, laboratory scientists, and graduate students. The editors have made a deliberate effort to select contributors with recognized expertise in citrus fruit processing. We extend our sincere gratitude to all the authors for their valuable contributions. Special thanks are also due to our readers and the editorial staff at Springer Nature for their support and insightful suggestions.

Dehradun, Uttarakhand, India — ***Shweta Suri***
Sonipat, Haryana, India — ***Anupama Singh***
Rennes, France — ***Kandi Sridhar***

Contents

Contributors

DIVYA ARORA • *Department of Food Engineering, National Institute of Food Technology Entrepreneurship and Management, Kundli, Haryana, India*

KANCHAN BHATT • *Department of Food Science and Technology, Dr. Yashwant Singh Parmar University of Horticulture and Forestry, Nauni, Solan, Himachal Pradesh, India*

CINTHIA BAÚ BETIM CAZARIN • *Universidade Estadual de Campinas, Campinas, São Paulo, Brazil*

RAJNI CHOPRA • *Department of Food Science and Technology, National Institute of Food Technology Entrepreneurship and Management, Kundli, Haryana, India*

PRIYANKA DUBEY • *Department of Biosciences, Integral University, Lucknow, Uttar Pradesh, India*

ALVINA FAROOQUI • *Department of Bioengineering, Integral University, Lucknow, Uttar Pradesh, India*

HAMID • *Department of Food Technology and Nutrition, Lovely Professional University, Phagwara, Punjab, India*

FOLAYEMI J. ISAAC-BAMGBOYE • *Department of Food Science and Technology, Abiola Ajimobi Technical University, Ibadan, Oyo State, Nigeria*

AFIYA JOHN • *Department of Food Production, Faculty of Food and Agriculture, The University of the West Indies, St Augustine Campus, St Augustine, Trinidad and Tobago*

BARJINDER PAL KAUR • *Department of Food Engineering, National Institute of Food Technology Entrepreneurship and Management, Kundli, Haryana, India*

SAMANDEEP KAUR • *School of Engineering and Technology, National Forensic Sciences University, Gandhinagar, Gujarat, India*

ANKUR KUMAR • *Central Instrumentation Laboratory, National Institute of Food Technology Entrepreneurship and Management (NIFTEM), Kundli, Haryana, India*

SANJANA KUMARI • *Department of Food Engineering, National Institute of Food Technology Entrepreneurship and Management, Kundli, Haryana, India*

SEERAT MANZOOR • *Department of Food Technology, Islamic University of Science and Technology, Awantipora, Jammu and Kashmir, India*

SAYANI MAVAI • *Department of Food Science and Technology, National Institute of Food Technology Entrepreneurship and Management, Kundli, Haryana, India*

FRANCESCO MENEGUZZO • *Institute of Bioeconomy, National Research Council of Italy, Florence, Italy*

PARMJIT S. PANESAR • *Sant Longowal Institute of Engineering and Technology, Longowal, Sangrur, Punjab, India*

JAN SABURA SHABEER • *Department of Food Technology, Islamic University of Science and Technology, Awantipora, Jammu and Kashmir, India*

OMINA SHOMIRZOEVA • *Department of Pharmacognosy and Organization of Pharmacy Economics, Avicenna Tajik State Medical University, Dushanbe, Tajikistan*

SIMRAN • *Central Instrumentation Laboratory, National Institute of Food Technology Entrepreneurship and Management (NIFTEM), Kundli, Haryana, India*

ANUPAMA SINGH • *Department of Food Engineering, National Institute of Food Technology Entrepreneurship and Management (NIFTEM), Sonipat, Haryana, India*

RAHUL SINGH • *Department of Bioengineering, Integral University, Lucknow, Uttar Pradesh, India*
VIKRANT SINGH • *Graphic Era (Deemed to be University), Dehradun, Uttarakhand, India*
SHWETA SURI • *School of Food Science, Washington State University, Pullman, USA; School of Health Science and Technology, University of Petroleum and Energy Studies (UPES), Dehradun, Uttarakhand, India*
ADRIANA GADIOLI TARONE • *Universidade Estadual de Campinas, Campinas, São Paulo, Brazil*
KAISER YOUNIS • *Department of Bioengineering, Integral University, Lucknow, India*
OWAIS YOUSUF • *Department of Food Technology, Islamic University of Science and Technology, Awantipora, Jammu and Kashmir, India*
FEDERICA ZABINI • *Institute of Bioeconomy, National Research Council of Italy, Florence, Italy*

Chapter 1

Overview of Citrus By-Products: Market Potential and Applications

Kanchan Bhatt, Hamid, and Shweta Suri

Abstract

Citrus fruits are among the most extensively utilized worldwide, recognized for their significant nutritional benefits. They are processed on an industrial scale, leading to the generation of substantial by-products, including peels, seeds, and pomace. These by-products contain a variety of secondary metabolites, such as sugars, coumarins, carotenoids, alkaloids, limonoids, flavonoids, phenolic acids, essential oils (EOs), and trace elements. Citrus peels are also a major source of pectin, dietary fiber that can be utilized as emulsifiers, fat substitutes, and gelling agents in jams, jellies, etc. Further, these by-products are also widely utilized in various segments of the food industry, besides their utilization in self-care products and pharmaceuticals. Hence, waste valorization can be a cost-effective solution to handle this citrus processing waste and to explore new opportunities for its wide utilization. Effective usage of citrus by-products will lead to waste minimization and support sustainable practices, leading to food security.

Key words Citrus by-products, Pomace, Peel, Waste valorization, Food applications

1 Introduction

Citrus fruits rank among the most extensively cultivated fruits globally. These fruits belong to the *Rutaceae* family and encompass a diverse range of varieties, including *Citrus reticulata* (mandarin), *Citrus sinensis* (orange), *Citrus tangerine* (tangerine), *Citrus limon* (lemon), *Citrus limetta* (sweet lime), *Citrus aurantifolia* (lime), and *Citrus paradisi* (grapefruit). Citrus group is recognized for its substantial nutritional value, attributed to vitamins (thiamine, riboflavin, niacin, and ascorbic acid), minerals, dietary fibers, and bioactive substances such as polyphenols, flavonoids, and carotenoids. Consequently, these bioactive compounds are sufficiently effective in offering numerous health advantages to people, including delaying aging, enhancing immunity, improving digestibility, and addressing various mental and cardiovascular disorders [1]. Additionally, the fruit's amazing flavor and massive production (more

Shweta Suri et al. (eds.), *Citrus Fruit Processing By-Products: Technologies and Basic Protocols*, Methods and Protocols in Food Science, https://doi.org/10.1007/978-1-0716-5088-2_1,

than 124.3 million tons globally) encourage its further processing into various beverages, intermediate moisture foods, and various other value-added products [2].

Every year, industrial processing of citrus fruits at mega and micro levels generates a massive amount of waste (10 million MT) globally [3]. These remnants are referred to as citrus by-products and consist of peels (50–55%), seeds (20–40%), pomace, and molasses. In addition to being huge, this waste is also perishable in nature (80–90% water) and classified as hazardous due to its adverse environmental effects. Waste valorization presents a cost-effective approach to addressing citrus waste, which encompasses a diverse array of secondary metabolites, including sugars, coumarins, carotenoids, alkaloids, limonoids, flavonoids, phenolic acids, essential oils, and trace elements [4]. Furthermore, the utilization of citrus by-products offers the potential to enhance environmental safety while providing advantages across various sectors, including food production, agriculture, pharmaceuticals, cosmetics, and packaging.

This highlights the latest applications & market potential of citrus processing waste as a possible supply of raw materials for the food and other nonfood industries. This utilization contributes to waste minimization and supports sustainable practices, leading to food security.

2 Materials

2.1 Categories of Citrus By-Products

Citrus by-products are mainly categorized as citrus peels, pomace, seeds, and molasses (Fig. 1). Citrus peels are the outermost layer of citrus fruits, divided into two layers: flavedo (brightly colored) and albedo (white colored). Citrus peels are a vital source of limonoids, flavonoids, essential oils, etc. They are used to extract pectin that is used as a gelling agent in the food sector, essential oils for flavorings and fragrances, and dietary fiber [5]. The fibrous residue containing peels, pulp, and seeds that remain after the juice extraction from citrus is collectively known as pomace. Besides, being a good source of polyphenols like hesperidin, naringin, narirutin, neoeriocitrin, rutin, ferulic acid, dried citrus pomace is a good source of sugars (30–40%), cellulose and hemicelluloses (13–17%), and pectin (14–25%) [6]. Citrus pomace consists of limonoids, flavonones, flavones, and various phenolic acids [7] that can be potentially transformed into various food additives like food colorants, flavorings, and sweeteners [8]. Furthermore, citrus seeds contain oils and various bioactive compounds. These oils can be used in cosmetics and biodiesel; however, seed cake can be used as a protein supplement for animals [4]. Lastly, citrus molasses is another by-product that is syrup-rich in sugars and can be used as a fermentation substrate for producing ethanol or as a component in animal feed [9].

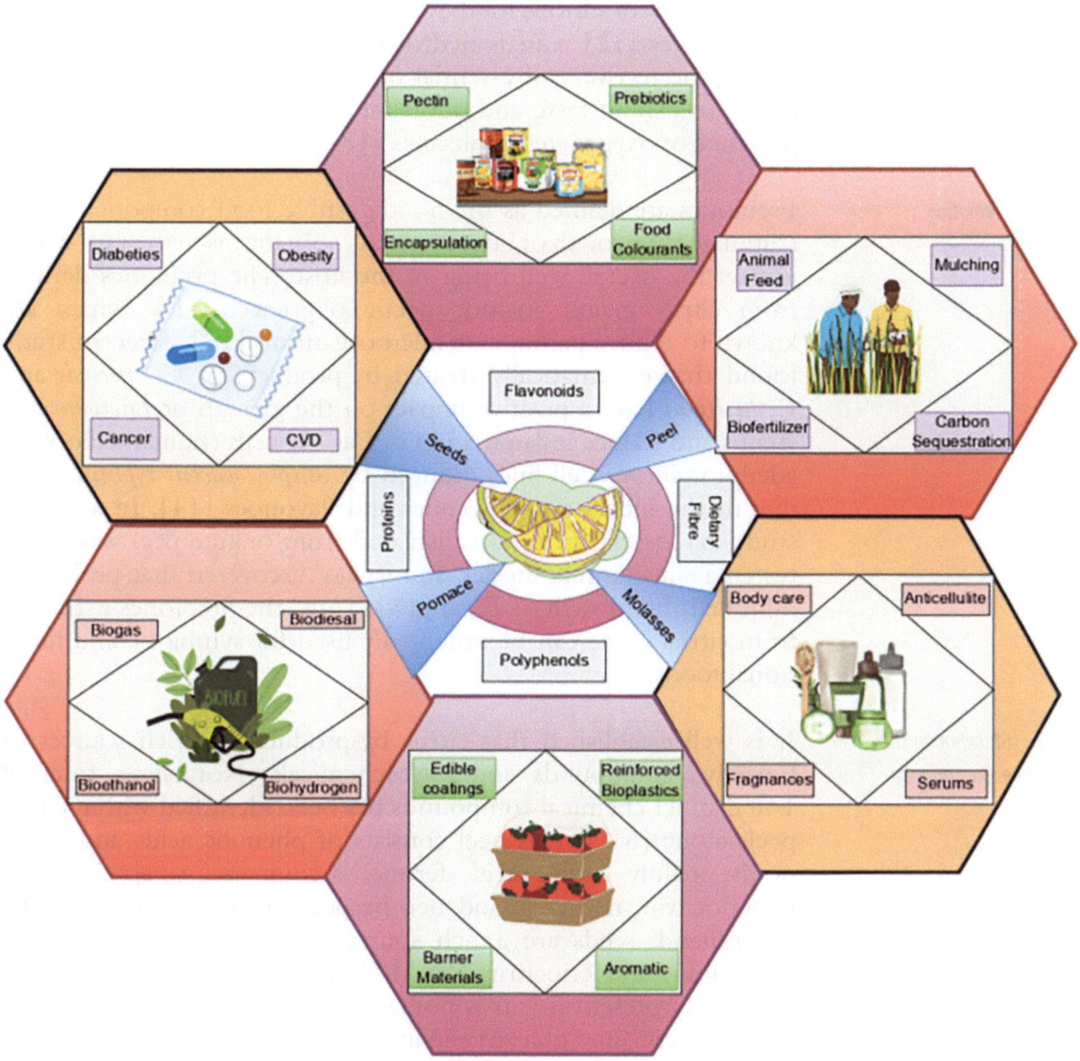

Fig. 1 Citrus by-products categories and potential applications

3 Methods

3.1 Application in Food Sector

3.1.1 Food Product Development

In several regions of India and Europe, citrus peels are incorporated into the preparation of custards, curries, sweet delicacies, and baked goods. This practice is employed to enhance natural flavor and improve color, thanks to the presence of various bioactive compounds, including carotenoids and essential oils [10]. Moreover, citrus peels are a major source of pectin (124 million tons annually), which is utilized as an emulsifier, fat substitute, gelling agent in jams, jellies [11], and various other foods like dairy products, sweets, soft drinks, frozen meals, and medications [5]. Citrus pomace, being a good source of dietary fiber, can be used to enrich the fiber content of various foods, for example, cakes, muffins, biscuits, and noodles [12]. Citrus hydrosols, obtained in minor amounts during the recovery of essential oils from by-products, have been found to exhibit taste and odor imparting capabilities due to the presence of oxygenated molecules [13].

3.1.2 Prebiotics

Prebiotics are defined as the nondigestible food components that stimulate the growth of beneficial microorganisms in the gut, which impact the overall well-being of the host. The prebiotics derived from citrus mainly include pectic oligosaccharides, which are known to reduce pathogenic microorganisms [13]. A recent study found that enzymatically treated by-products of *C. sinensis* and *C. latifolia* have a positive impact on the growth of *Lactobacillus* strains and reduce inflammation in $CaCO_2$ cells (human colorectal adenocarcinoma cell line) induced by *Staphylococcus typhimurium* due to the increase in aglycones and flavonoes [14]. In a similar study, oligosaccharides were isolated from orange peel waste and tested against a milk model, and it was discovered that probiotics inhibited the growth of *E.coli* [15]. Thus, the prebiotics extracted from citrus waste can be potentially used for symbiotic and functional foods.

3.1.3 Storehouse of Bioactive Compounds

It is well established that citrus by-products are rich sources of bioactive compounds and phytochemicals. Notably, a total of 140 distinct chemical compounds has been identified within citrus peels alone [5]. Citrus peel consists of phenolic acids and flavonoids, mainly caffeic acid, ferulic, p-coumaric, sinapinic acids, neoeriocitrin, naringin, and neo-hesperidin, respectively. On the other hand, seeds are a rich source of proteins, fats, minerals, hesperidin, and eriocitrin [16, 17]. Furthermore, dietary fibers are mostly present in citrus pomace and are of two types: soluble (gums, pectin, and glucan) and insoluble (cellulose, hemicelluloses, and lignin) [13]. Bioactive chemicals such as phenolic acids have been linked to reduced risk of cancer, aging, diabetes, and other lifestyle diseases. However, flavonoids are renowned for their anti-inflammatory and antiviral effects. Thus, the high concentration of dietary fibers and bioactive substances in citrus by-products highlights its potential as a significant ingredient in the development of functional foods aimed at enhancing health and preventing lifestyle illnesses [16].

3.1.4 Encapsulating Agent

Encapsulation is the process by which active ingredients are enclosed within a protective coating to enhance their stability, functionality, and controlled release. These are mainly used in the pharmaceutical and nutraceutical industries to prevent active ingredients from various types of environmental impacts, as they are sensitive molecules and can be denatured in harsh conditions [18]. According to research, orange juice by-products have been identified as a potential encapsulating material for pomegranate peel extract due to their high pectin content, which may protect sensitive bioactive components better [19]. Microcapsules were developed using citrus pectin and sodium alginate using the spray drying technique, and it was concluded that powder dried at 100 °C showed maximum retention efficiency, antioxidant activity, and antimicrobial properties against *E. coli* bacteria [20]. Encapsulation technology is critical in addressing issues such as stability, potency, and targeted distribution of sensitive and active substances, which drives innovation across industries.

3.2 Application in Non-Food Sector

3.2.1 Packaging Material

Researchers are interested in biopolymer by-products and advanced packaging techniques like active and intelligent packaging because they may help mitigate the detrimental impact of disposing of plastic on the environment and reliance on fossil fuels. Citrus by-products are a proven asset in this arena because of their high carbohydrate content. In a recent study conducted by Khalil et al. [21], it was found that the edible films made up of pectin extracted from grapefruit peel were more efficient than commercial pectin extracted from citrus. Further, these films were known to have more thermal stability and cost-effectiveness. Incorporation of essential oil to edible films made from citrus waste enhances the moisture stability of the packaging material, besides imparting it antimicrobial properties [22]. Whey protein isolate containing citrus peel oil emulsions and nano-emulsions was found to have superior water barrier, antimicrobial, and antibacterial properties. This active packaging system was studied and recommended by [23] for active food packaging.

3.2.2 Source of Biofuels

Usage of biomass to produce energy rather than nonrenewable fuels has garnered international attention in recent years. Various investigations concluded that citrus peel waste can act as a promising biofuel as it contains numerous organic compounds like fermentable sugars and low lignin content [13]. Technologies like anaerobic digestion and microbial fermentation are used to produce biofuels like bioethanol and biodiesel, as they provide higher yield and minimum emissions. Pyrolysis and torrefaction methods can be used for the production of bio-hydrogen, bio--chars, and bio-tars from lemon and orange peels [2]. In a study conducted by [24], orange waste was used as bio-refinery feedstock to produce pectin, ethanol, and biogas. Orange waste (peel and pulp) was subjected to sulphuric acid treatment, and pectin was extracted from the hydrolysate. The liquor was further fermented to produce ethanol, and the remnant solid was subjected to anaerobic digestion to produce biogas. The citrus waste valorization as biofuels aims toward carbon neutrality, reduced greenhouse emissions, decreased air pollution, with increased energy security.

3.2.3 Bio-fertilizer

Bio-fertilizers are different from conventional fertilizers as they contain living microorganisms (bacteria, fungi, and algae) that colonize the rhizosphere and increase the flow of nutrients to the plant. It is noted that bio-fertilizers are eco-friendly, cost-effective, and enhance soil fertility and texture [25]. In a pilot study conducted by Lee et al. (2019) [26], by-products from *Citrus reticulata* were found to be efficient in removing cadmium from soil by the presence of high levels of citric acid and enhancing the nutrient concentrations by immobilization. Fermented orange peel liquor was used to treat barley varieties in a study as a biofertilizer. Absence of hemolytic activity and coliforms, along with an increase in protein and chlorophyll content, was confirmed from various assays in this study [27].

3.2.4 Self-Care Products

The market for natural cosmetics has grown in the past few years; however, it is necessary to acknowledge the excellent research that has been done and the many publications that highlight the significance of bioactive components. Citrus by-products fulfill the demand of the cosmetic industry as they are a storehouse of various antioxidant-rich compounds like vitamin C and E, which are effective in the prevention of ageing of skin. They also consist of flavonoids like naringin, hesperidin, quercetin, and rutin, which help in achieving wrinkle-free and radiant skin [13]. Additionally, essential oils present in citrus by-products (Table 1) are known to inhibit melanin-producing enzyme tyrosinase, thus securing a position in skin whitening cosmetics [28]. In a study conducted on California orange peel extract, it was found that when the extract (2%) was used in facial cream, it reduced the melanin content by 17.33% in 1 month without any case of allergy [29].

3.2.5 Pharmaceuticals

Citrus fruits and their by-products have been used for various medicinal purposes around the world since ancient times. For instance, in Japan, the dried peels of ripe citrus fruits are used as crude drugs known as "Chimpi" and "Touhi." A similar practice of utilizing dried peels can be found in Chinese herbal medicine, where they are used to treat bronchitis and asthma [5]. A study revealed the effects of aqueous citrus peel extracts, butanol, and hesperidin against a rat model of ulcer and hepatotoxicity. Further, it was found that hesperidin exerted the best hepatoprotective, antibacterial, and anti-inflammatory effect, followed by butanol and then aqueous citrus peel extracts [30]. Pectin from citrus by-products is found to have prime importance in the pharmaceutical industry, as it is used as a gelling, thickening, and hemostatic agent, and it is also used with other colloids to control drug release. In another research, it was found that the peels of Kawachi Bankan (*Citrus kawachinensis*) consist of neuroprotective components, namely, auraptene and 3,5,6,7,8,3′,4′-heptamethoxy flavones, and both promote neuron functioning and its protection [31].

Table 1
Utilization and health benefits of citrus by-products in different industries

Citrus by-products	Area of applications	Products	Bioactive compounds	Method of extraction	Health benefits	Results	References
Peel	Food sector	Prebiotic	Dietary fiber (soluble and insoluble) Oligosaccharides	Ultrasound-assisted extraction (9 min at 25 °C) plus Drying (40 °C)	Better gut health	Increased CFU count both for *Lactobecillus returi* and *Bifidibacterium* Decrease in *E.coli*	[14]
Pomace		Biscuits	Dietary fiber	Juice extraction followed by Drying (60 °C for 24 h), and Grinding into powder form	Good source of vitamin C and carotenoids	High amount of fiber (60.12%) High water holding capacity Low oil holding capacity Reduction in 'L' and 'b' values and increase in 'a' value	[12]
Seed		Biscuits	Vitamins (A, B, D, E, and K) Fat Minerals (Ca, Mg, K, Na, P, Mn, Fe, and Zn w)	Removal of seeds followed by Cleaning, Dehulling, Drying, and Grinding to obtain Orange seed powder	Enhanced antioxidant activity	Increased fiber content Increased amount of vitamins in biscuits	[17]
Peel	Nonfood sector	Cosmetics Pharmaceutical	Sinensetin, 4,5,6,7-tetramethoxyflavone, Nobiletin, 3,3',4',5,6,7-hexamethoxyflavone, Narirutin	Peeling of Newhall navel orange followed by Drying, and Vaccum evaporation to	Antioxidant activity Anti-inflammatory activity	Ethyl acetate extract of orange peel was found to have the best biological effects Inhibition of a wide spectrum of	[28]

(continued)

Table 1 (continued)

Citrus by-products	Area of applications	Products	Bioactive compounds	Method of extraction	Health benefits	Results	References
				obtain Orange peel extracts		microorganisms *E. coli*—24.50 m *B. subtilis*—30.36 m *S. aureus*—11.75 mm *X. citri* subsp. Citri—19.01 mm Tyronise inhibitory activity—IC50 value of 108.24 μg mL^{-1}	
Pomace		Neutraceuticals	Flavones Flavanones Phenolic acids Limonoids	Raw orange and lemon pomace was passed through Ultrasound-assisted extraction (Room temperature) followed by Vaccum evaporation to obtain Pomace extracts	Antioxidant activity Anti-inflammatory activity	Pomace extracts were added to simulated gastric fluid and the effect on human colorectal adenocarcinoma Caco-2 cell line was evaluated. It was found that Both extracts antioxidant activity by reducing the rate of increase of LPS-dependent ROS levels in Caco-2 cells Lemon extract	[7]

					oppressed the effect of pro-inflammatory cytokines (L-1β, IL-6, IL-8, and TNF-α)	
Seed	Packaging film	Citric acid Polyphenols Flavonoids Tocopherol Ascorbic acid	Grapefruit seeds were washed, dried, ground to powder form	Antioxidant activity Antimicrobial activity	Graprfruit seed extract, when applied directly or indirectly by adding to the polymer matrix enhances the following: Water solubility Swelling properties Thermal stability Decreases mechanical strength	[22]
Molasses	Livestock production	Alkaloids Flavonoids Glycosides Saponin Steroids Phenols Terpenoides Tannin Antraquinones	Eggs + lime juice (1 litre) + molasses (500 g) were left covered for 10 days (27 °C and 61% humidity) followed by blending of solution	Good source of minerals and phytochemicals Natural antibiotic	Egg lime molasses has potential to help physicological processes in human and animals Help to prevent chronic diseases	[9]

3.3 Market Scenario of Utilization of Citrus By-Products: An International Trend

Citrus is the most extensively grown crop in the world, and demand for it is rising rapidly. Apart from its famous refreshing juice, its by-products—such as peels, seeds, and pomace—are also valuable assets to various industries like food, medicine, cosmetics, and nutraceuticals because they include dietary fiber, pectins, flavonoids, phenols, minerals, vitamins, and other nutrients. These by-products represent approximately 50% of the total fruit's weight; thus, waste valorization can help to manage them efficiently. The market for citrus extracts is dominated by the Asia Pacific area because of its high level of production. Nowadays, citrus peel-based teas, chocolates, essential oils, and their powder for cosmetic purposes rule the market [28]. A report published in fact. MR states that the citrus peel market in 2022 was valued US $ 7.59 billion and projected to reach US $ 11.84 billion by 2032 at a CAGR of 4.5% [32]. Nonetheless, the market for citrus seeds is expected to continue expanding due to ongoing research and development in its application. This expansion demonstrates the unrealized potential of citrus by-products in developing globally sustainable and profitable solutions.

3.4 Future Research on the Commercialization of Citrus By-Product Constituents

It is very evident from the information collected that citrus by-products have the utmost potential to bring a sustainable change to the planet. The bio-actives found in citrus by-products must be mindfully extracted using advanced techniques to retain their stability and bio-availability. Additionally, in order to attain higher yields, environment-friendly techniques like supercritical fluid extraction, pressurized liquid extraction, and pre-treatments using deep eutectic solvents must be used. Advanced research into therapeutic applications with more in vitro and in vivo trials is essential to unlock their health benefits. High-demand areas that require research on a priority basis, like bio-plastics, biodegradable packaging, and bio-energy production, must be offered more funding by different research organizations worldwide. Some challenges persist in due course of utilization, like the seasonal nature of citrus fruits, improper waste disposal due to its volume, and the high cost of processing. Thus, an interdisciplinary approach, collaborative effort, and scalable studies are required to understand the concept of waste valorization, leading to consumer-focused product development essential for ensuring feasibility and large-scale adoption of sustainable practices.

4 Notes

Citrus is the most widely available fruit for processing purposes, which generates huge by-products in the form of flavedo, albedo, pomace, seeds, etc. These by-products also have high nutritional value with bioactive compounds that can be significantly utilized in the development of various value-added products with modified food properties and enhanced nutritional value. These by-products have good potential to be utilized in various segments of the food industry, for example, in new product development, in dairy products, in beverages, besides use in other segments in self-care products, in pharmaceuticals, etc. However, for commercialization in the food industry, this feasibility in combination with novel techniques upholds new future opportunities by large-scale adoption of sustainable practices.

References

1. Bora H, Kamle M, Mahato DK, Tiwari P, Kumar P (2020) Citrus essential oils (CEOs) and their applications in food: an overview. Plants 9(3):357
2. Mahato N, Sharma K, Sinha M, Baral ER, Koteswararao R, Dhyani A, Cho MH, Cho S (2020) Bio-sorbents, industrially important chemicals and novel materials from citrus processing waste as a sustainable and renewable bioresource: a review. J Adv Res 23:61–82
3. Suri S, Singh A, Nema PK (2022) Current applications of citrus fruit processing waste: a scientific outlook. Appl Food Res 2(1):100050
4. Maqbool Z, Khalid W, Atiq HT, Koraqi H, Javaid Z, Alhag SK, Al-Shuraym LA, Bader DMD, Almarzuq M, Afifi M, Al-Farga A (2023) Citrus waste as source of bioactive compounds: extraction and utilization in health and food industry. Molecules 28(4):1636
5. Liu N, Li X, Zhao P, Zhang X, Qiao O, Huang L, Gao W (2021) A review of chemical constituents and health-promoting effects of citrus peels. Food Chem 365:130585
6. Caballero S, Li YO, Mc Clements DJ, Davidov-Pardo G (2021) Encapsulation and delivery of bioactive citrus pomace polyphenols: a review. Crit Rev Food Sci Nutr 62(29):8028–8044
7. Ingegneri M, Braghini MR, Piccione M, De Stefanis C, Mandrone M, Chiocchio I, Poli F, Imbesi M, Alisi A, Smeriglio A, Trombetta D (2024) Citrus pomace as a source of plant complexes to be used in the nutraceutical field of intestinal inflammation. Antioxidants 13(7): 869
8. Wedamulla NE, Fan M, Choi YJ, Kim EK (2022) Citrus peel as a renewable bioresource: transforming waste to food additives. J Funct Foods 95:105163
9. Akintunde AO, Ndubuisi-Ogbonna LC, Olorunfemi OA, Ladele MM, Shobo BA, Adewole SA, Akinboye OE (2023) Nutritional and ethno-medicinal potentials of egg-lime-molasses mixture in livestock production. Asian J Dairy Food Res 42(3):298–306
10. Xiao L, Ye F, Zhou Y, Zhao G (2021) Utilization of pomelo peels to manufacture value-added products: a review. Food Chem 351: 12924. https://doi.org/10.1016/j.foodchem.2021.129247
11. Huang JY, Liao JS, Qi JR, Jiang WX, Yang XQ (2020) Structural and physicochemical properties of pectin-rich dietary fiber prepared from citrus peel. Food Hydrocoll. https://doi.org/10.1016/j.foodhyd.2020.106140
12. Hussain A, Kauser T, Aslam J, Quddoos MY, Ali A, Kauser S, Gorsi FI (2023) Comparison of different techno-functional properties of raw lemon pomace and lemon pomace powder, and development of nutritional biscuits by incorporation of lemon pomace powder. J Sustain Agric 38(1):76–192
13. Panwar D, Saini A, Panesar PS, Chopra HK (2021) Unraveling the scientific perspectives of citrus by-products utilization: progress towards circular economy. Trends Food Sci Technol 111:549–562
14. de Paula Menezes Barbosa P, Ruviaro AR, Martins IM, Macedo JA, LaPointe G, Macedo GA (2020) Effect of enzymatic treatment of citrus by-products on bacterial growth, adhesion and cytokine production by Caco-2 cells. Food Funct 11(10):8996–9009
15. Zafar J, Iahtisham-Ul-Haq NGA, Ramniwas S, Mugabi R, Ali Alharbi S, Ansari MJ (2024) Studies on the growth of *Lactobacillus reuteri*, *Bifidobacterium* and *Escherichia coli* as affected by prebiotic extracted from citrus peel. Int J Food Prop 27(1):783–798
16. Kaur S, Panesar PS, Chopra HK (2023) Citrus processing by-products: an overlooked

repository of bioactive compounds. Crit Rev Food Sci Nutr 63(1):67–86

17. Hussain A, Laaraj S, Kausar T, Tikent A, Azzouzi H, Kauser S, Ul An Q, Iqbal A, Akram S, Nisar R, Najam A, Fatima H, Yakub S, Elfazazi K (2023) Food application of orange seed powder through incorporation in wheat flour to boost vitamin and mineral profiles of formulated biscuits. Int J Food Sci 2023(1):6654250
18. Comunian TA, Silva MP, Souza CJ (2021) The use of food by-products as a novel for functional foods: their use as ingredients and for the encapsulation process. Trends Food Sci Technol 108:269–280
19. Kaderides K, Mourtzinos I, Goula AM (2020) Stability of pomegranate peel polyphenols encapsulated in orange juice industry by-product and their incorporation in cookies. Food Chem 310:125849
20. Sun X, Cameron RG, Bai J (2020) Effect of spray-drying temperature on physicochemical, antioxidant and antimicrobial properties of pectin/sodium alginate microencapsulated carvacrol. Food Hydrocoll 100:105420
21. Khalil RK, Sharaby MR, Abdelrahim DS (2023) Novel active edible food packaging films based entirely on citrus peel wastes. Food Hydrocoll 134:107961
22. Roy S, Zhang W, Biswas D, Ramakrishnan R, Rhim JW (2023) Grapefruit seed extract-added functional films and coating for active packaging applications: a review. Molecules 28(2):730
23. Amjadi S, Almasi H, Ghadertaj A, Mehryar L (2021) Whey protein isolate-based films incorporated with nanoemulsions of orange peel (*Citrus sinensis*) essential oil: preparation and characterization. J Food Process Preserv 45(2):e15196
24. Vaez S, Karimi K, Mirmohamadsadeghi S, Jeihanipour A (2021) An optimal biorefinery development for pectin and biofuels production from orange wastes without enzyme consumption. Process Saf Environ Prot 152:513–526
25. Abobatta WF (2020) Biofertilizers and citrus cultivation. MOJ Ecol Environ Sci 5(4): 171–176
26. Lee HH, Heo DY, Han HR, Park YL, Ruangcharus C, Kim SU, Seo DC, Oh TK, Hong CO (2019) Evaluation of the effects of Mandarin (*Citrus reticulate*) by-products containing citric acid on immobilization of cadmium in arable soils. Appl Biol Chem 62(1): 45. https://doi.org/10.1186/s13765-019-0452-2
27. Elsakhawy T, Alshaal T, Elakhdar A, El-Akhdar I (2024) From waste to worth: using fermented orange pomace in sustainable feed production. Plants 13(22):3191
28. Guo C, Shan Y, Yang Z, Zhang L, Ling W, Liang Y, Ouyang Z, Zhong B, Zhang J (2020) Chemical composition, antioxidant, antibacterial, and tyrosinase inhibition activity of extracts from Newhall navel orange (*Citrus sinensis* Osbeck cv. Newhall) peel. J Sci Food Agric 100(6):2664–2674
29. Ben Hsouna A, Sadaka C, Generalic Mekinic I, Garzoli S, Svarc-Gajic J, Rodrigues F, Morais S, Moreira MM, Ferreira E, Spigno G, Borjan B, Akacha BB, Saad RB, Matos CD, Mnif W (2023) The chemical variability, nutraceutical value, and food-industry and cosmetic applications of citrus plants: a critical review. Antioxidants 12(2):481
30. Abou Baker DH, Ibrahim BM, Abdel-Latif Y, Hassan NS, Hassan EM, El Gengaihi S (2022) Biochemical and pharmacological prospects of *Citrus sinensis* peel. Heliyon 8(8):e09979
31. Furukawa Y, Okuyama S, Amakura Y, Sawamoto A, Nakajima M, Yoshimura M, Yoshida T (2021) Isolation and characterization of neuroprotective components from citrus peel and their application as functional food. Chem Pharm Bull 69(1):2–10
32. Fact MR (2023) Citrus peel market: global industry analysis, size, share, growth, trends, and forecast, 2022–2032. Retrieved from https://www.factmr.com

Chapter 2

Extraction of Bioactive Compounds from Citrus By-Products

Francesco Meneguzzo and Federica Zabini

Abstract

Citrus by-products, resulting from juice squeezing in an amount up to 50% of the fruit's fresh weight, are currently discarded or underutilized, with most of the valuable bioactive compounds lost to animal feeding or energy generation. This represents a serious economic and environmental burden. Several conventional and novel green extraction methods of citrus by-products are discussed along with their advantages and drawbacks, and the potentially most promising method is framed within the broader process chain leading to end products in the form of dry extracts, which are suitable for pharmaceutical and nutraceutical industries.

Key words Citrus by-products, Bioactive compounds, Barriers, Green extraction, Pharmaceutics, Nutraceutics

1 Introduction

Citrus by-products, a mix of peels, seeds, and pulp residues resulting from juice squeezing and production of jams, concentrated formulas, or pastes, as well as sometimes the extraction of essential oils (EOs), which will be referred to as citrus peel waste (CPW), constitute a substantial fraction of the fresh weight of citrus fruit. Every year, it is estimated that worldwide about 15 million tons of CPW are produced [1], representing 40–50% of the fruit's fresh weight in the case of oranges [2].

More in detail, CPW peels are composed of flavedo and albedo, the former being the outside colorful part of the peel that contains the oil sacs, while the latter is the white interior part of the peel, rich in pectin. Peels are also rich in sugars and have a high concentration of D-limonene, an essential oil and powerful antimicrobial compound [1]. Pulp residue consists of membranes that contain the juice, and seeds mainly contain lipids, crude protein, and fiber. CPW is endowed with several active compounds with powerful bioactive activities, such as antioxidant and antimicrobial. Pectin, vitamin C, carotenoids, flavonoids, and essential oils are the main

Shweta Suri et al. (eds.), *Citrus Fruit Processing By-Products: Technologies and Basic Protocols*, Methods and Protocols in Food Science, https://doi.org/10.1007/978-1-0716-5088-2_2,

bioactive compounds available in the CPW from any citrus varieties [1].

Yusuf et al. produced a comprehensive assessment and discussion of the barriers hindering the sustainable exploitation of CPW toward the circular economy target [3]. For the extraction of bioactive compounds from CPW, they found that the main barriers, in descending order of numerosity, were technological, economic, informational, and regulatory. Any strategy devised for the extraction of CPW bioactive compounds should be assessed based on its potential ability to overcome such barriers.

2 Materials

Economic sustainabiliy in extracting bioactive compounds from CPW hinges on minimizing capital expenditure for industrial-scale plants and reducing operating costs during processing. Lower operating costs can be achieved with the extraction and exploitation of integral phytocomplexes rather than individual substances, due to the high purification costs [3]. As well, the application fields of the end products should be as valuable as possible, which, in the case of bioactive compounds, turn out to be nutraceuticals and pharmaceuticals, as shown in Fig. 1.

Overcoming technological barriers, in the case of the extraction of CPW bioactive compounds, amounts to developing techniques ready for scale-up to the industrial production capacity, while avoiding increasingly unacceptable and costly adverse environmental impacts [3]. This requirement translates into the compliance of the technique with the consolidated and recently updated and extended green extraction principles [5], including the use of water as the only solvent, the efficient use of biomass, the predictability, and scalability. However, the end products should also be safe, stable, and effective, a precondition for the latter attribute being sufficient bioavailability, which is particularly challenging for poorly water-soluble flavonoids, such as hesperidin, which is one of the most powerful and widespread among citrus flavonoids.

The information barriers relate to proper communication to farmers, entrepreneurs, investors, and consumers about the potential health, ecological, and economic advantages resulting from CPW exploitation [3]. Overcoming this barrier may require scholars to get more deeply involved in entrepreneurial and technological education and training.

Finally, the regulatory barrier is even more subtle, as regulations affect both the very possibility of using CPW, often disregarded as a general waste, and challenge the marketing of the extracted bioactive compounds due to safety concerns. The European Union's novel food regulation [6] requires pre-market authorization not only for ingredients that were not consumed to a

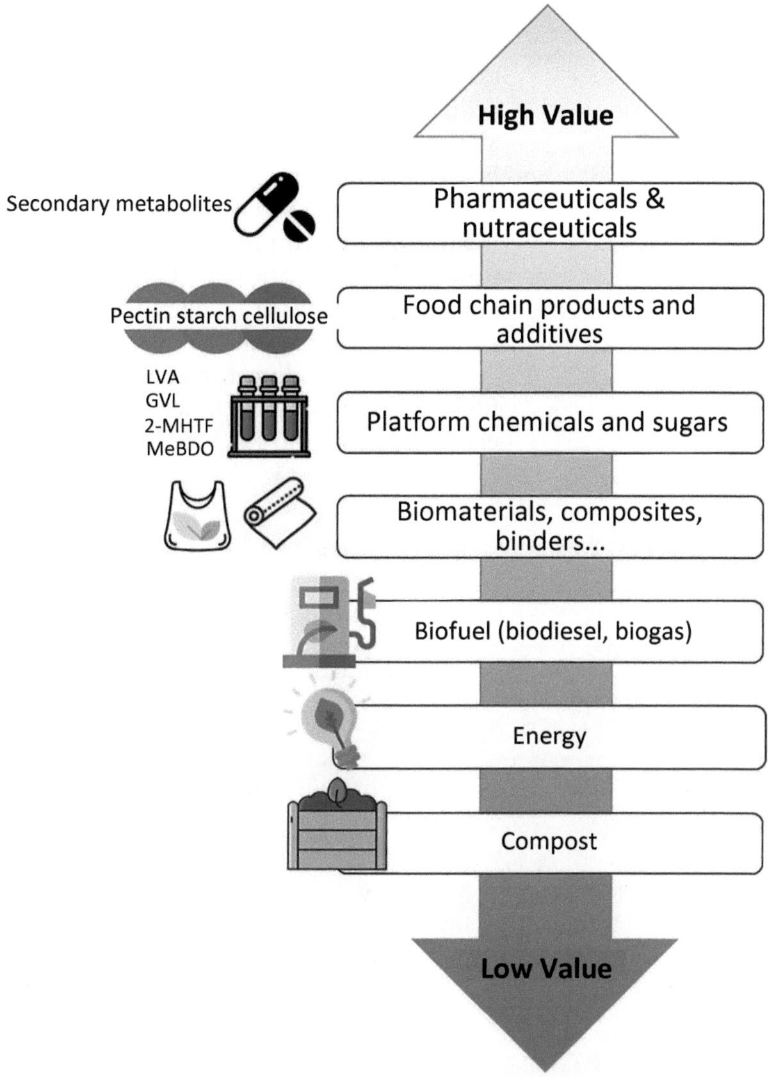

Fig. 1 Value hierarchy for products derived from citrus by-products. (Reproduced from Meneguzzo and Zabini [4])

significant degree in the EU before 15 May 1997, but also for ingredients produced using new processes or technologies. Applicants must demonstrate safety, including where the production method results in a new or intentionally modified primary molecular structure.

Hybridization of the ancestral species of the citrus genus, namely mandarin, citron, and pomelo, gave rise to the wide range of current varieties, which are produced in the following order of overall quantity: oranges > tangerines, mandarins, clementines, satsumas > lemon and limes > minor citrus fruits including bergamot, citron and chinotto > grapefruits (including pomelo), totaling about 158 Mton in 2019 [1]. CWP incorporates a wide range of phytochemicals, such as the ubiquitous and abundant heteropolysaccharide pectin, polyphenols and especially flavonoids, phenolic acids, carotenoids, limonoids and EOs, with the respective amounts also changing with growing location and harvest time. All the bioactive compounds contribute to the characteristic aroma, flavor, and therapeutic potential of citrus peels [7, 8]. Table 1 shows the main bioactive compounds extracted from CPW, along with their potential therapeutic activities. Among flavonoids, only glycosides are present; however, generally the respective aglycone metabolites show similar activities.

Notwithstanding the challenging barriers, CPW-derived purified products have been actively introduced on the market, especially the flavonoid hesperidin also following its alleged benefits in

Table 1
Main Citrus Peel Waste bioactive compounds and their therapeutic value

Compound	Type	Main citrus varieties	Main therapeutic activities	References
Pectin	Heteropolysaccharide	All	Prebiotic; cholesterol-lowering; glycemic control; anti-constipation; metal chelation	[9, 10]
			Anti-inflammatory	[10, 11]
			Immunomodulatory; anticoagulant; gastroprotective	[11]
Hesperidin	Flavonoid (flavanone)	Orange	Antioxidant; antidiabetic; antihyperglycemic; antihyperlipidemic	[12–14]
			Antiviral	[13, 15]
			Anticancer	[13, 16, 17]
			Anti-inflammatory; antibacterial; anti-UV; anti-aging and skin barrier enhancement; wound healing	[13, 17]
			Neuroprotective; anti-neurodegenerative; anxiolytic, and antidepressant	[18]
Naringin	Flavonoid (flavanone)	Orange; grapefruit	Antihypertensive; cardioprotective; protective of the vascular endothelium	[19–21]
Eriocitrin	Flavonoid (flavanone)	Lemon; lime	Antidiabetic; antioxidant; anticancer; anti-allergic; anti-inflammatory; hepato-protective	[14, 22]
Limonene	Essential oil (monoterpene)	All	Anti-inflammatory; antioxidant; antiproliferative; anxiolytic, and antidepressant	[23]
			Neuroprotective; anti-neurodegenerative	[24]
			Immunostimulant	[25, 26]

counteracting the SARS-CoV-2 infection [15]. However, the conventional extraction methods used are detrimental to the environment and time-consuming and most flavonoids, including hesperidin, show both low solubility and extractability in solvents and poor bioavailability [3].

3 Methods

3.1 Conventional Extraction Methods

(a) *Maceration*
Maceration is the easiest method used for the recovery of bioactive compounds from plants, fruits, and fruit by-products, using a lengthy process comprising several steps [27]. Cold maceration involves placing fresh, ground biomass in a mixture with a solvent at room temperature with gentle agitation for an extended time (8–24 h). Then the extract is filtered, concentrated, and dried. Dar et al. observed a maximum extraction yield from bitter orange peels (*Citrus arantium*) of 29.28% using aqueous ethanol as a solvent [28]. In the case of kinnow mandarin peel, the highest maceration-based extraction yield (18.46%) was found with 80% ethanol at 35–40 °C for 24 h [29]. In another case, the extraction yields total phenolics from peels and pulp of mandarin, lemon, and orange using 80% ethanol at 70 °C for 3 h were at the levels of 169.54 mgGAE/100 g, 178.90 mgGAE/100 g and 61.22 mgGAE/100 g, respectively [30].

(b) *Soxhlet*
Soxhlet is a solid–liquid method to extract bioactive compounds from plants and fruits, based on the sequential evaporation and condensation of the selected solvent, around its boiling temperature, into the sample. Different types of solvents require different temperatures for the optimization of the extraction process: for methanol and ethanol, a temperature below 80 °C, while for deionized water, a temperature around 100 °C [31]. Jagannath et al. applied Soxhlet for the valorization of fresh lemon peels and obtained an aqueous extract with a total phenolic content of 68.36 mg GAE/100 g of extract and 39.55 mg Vitamin C/100 g of extract after a 65-min process at 65 °C [32].

(c) *Hydrodistillation*
Hydrodistillation is a traditional, easy, and cheap method to obtain volatile compounds, in particular EOs, from plants and fruit by-products. Hydrodistillation uses steam or water to extract the volatile compounds through a condenser, followed by the cooling of vapor and the return to the liquid form, before separation of water and volatiles into the collection flask. An EOs extraction yield of 4.5% was achieved after the

treatment of citron peels for 180 min at a temperature of 100 °C and pressure of 1 bar [33]. Hydrodistillation allows sustainable and operationally simple extraction for both fresh and dried citrus peel samples, showing a remarkable extraction yield of all the characteristic volatile compounds [34].

3.2 Novel Green Extraction Methods

(a) *Ultrasound Assisted Extraction*

Ultrasound-assisted extraction (UAE) methods are the most common across non-conventional green methods to obtain phenolic and volatile compounds from citrus by-products. UAE is a solid–liquid extraction method, based on the phenomenon of acoustic cavitation, i.e., generating vapor bubbles within a liquid, the implosion of which produces the fragmentation of molecules, cell walls, increases the exchange surface, and promotes the release of bioactive compounds [35].

Cavitation in liquid media is a multiphase phenomenon consisting of the generation, growth, and quasi-adiabatic collapse of vapor-filled bubbles under an oscillating pressure field, resulting in pressure shockwaves (up to 1000 bar), hydraulic jets, extreme local temperatures (up to thousands of K), and the generation of free radicals [36, 37].

The main parameters affecting the recovery efficiency are particle size, temperature, time, solvent type, and ultrasonic power and frequency [38]. Using water as the only solvent, the extraction rate of phenolic acids (*p*-coumaric, caffeic, and chlorogenic) and hesperidin from *Citrus limon* L. pomace (peel, membranes, and seeds) significantly increased with increasing temperature and decreasing particle size, [39]. In the case of pectin recovery, the extraction rate negatively correlated with the liquid to solid ratio, with a significant decrease in pectin yield with increasing solvent volume [40]. Several studies were performed to validate and optimize UAE for the extraction of polyphenols, pectin, essential oils, and carotenoids [41].

(b) *Pulsed Electric Field*

Pulsed electric field (PEF) is an emerging nonthermal method for the extraction of bioactive compounds, based on the application of microsecond pulses of high intensity electric fields that induce a nonreversible electroporation in cell membranes, with the cell disintegration index used to evaluate the appropriate PEF treatment conditions [42]. The intensity of the electric field, affecting the structural characteristics of the target material, is the most important determinant of the extraction yield, followed by the frequency and amplitude of the pulses, the wave shape, and the exposure time [43]. The temperature of the solvent in the treatment chamber is a decisive parameter for the diffusion of electrical pulses, with the process temperatures of extractive processes usually below 90 °C, both to ensure the preservation of functional compounds and to avoid the reduction in water viscosity that occurs at high temperatures.

PEF is also a technique applied in the pre-treatment of biomass to increase the extraction rate in subsequent processes. Hwang et al. compared conventional flavonoids extraction of Citrus *unshiu* with PEF followed by Subcritical Water Extraction (SWE), with an increase of hesperidin concentration by about 22.4, 2.1 and 1.2 times in comparison with hot water, methanol, and SWE, respectively [44]. Moreover, Luengo et al. found an increase by 20–159% of the extraction yield of the most abundant polyphenols from orange peels, like naringin and hesperidin, applying PEF before conventional pressing [45].

As a pre-treatment of orange peels, PEF also showed advantages in increasing the EOs extraction yield, with a 33% increase in limonene extraction when combined with ethanol [46]. Notably, the selectivity of the extracted aromatic compounds depended on the type of solvent used and the relative affinity to intracellular compounds that solubilize easily following treatment with PEF.

(c) *Microwave Assisted Extraction*

Microwave-assisted extraction (MAE) takes advantage of the dielectric properties of the plant matrix, sustaining the interaction of the polar molecules of the sample with the solvent, which in turn causes rapid heating and rupture of plant cell walls. MAE operates through two mechanisms of heat generation: dipolar rotation, with water evaporation; ionic conduction through the resistance of the solution to ion flow, producing friction and heat [35].

Combined MAE with solvent (ethanol) extraction showed superior flavonoid recovery from orange peels compared with conventional methods, operating at low temperatures and in a short time [47]. However, optimization of the microwave power and the exposure time to microwaves is critical to avoid degradation of phenolic and thermolabile compounds [48]. MAE, combined with steam diffusion, was used to increase the extraction yield of EOs from citrus by-products, such as D-limonene, reducing the processing time by about 3–20 times and achieving a better score of aroma sensory profile [49].

(d) *Enzyme Assisted Extraction*

Enzyme assisted extraction (EAE) can be useful in the case of particularly tight molecular bonds that are difficult to cleave. The enzyme used hydrolyzes the cell walls, releasing the target compounds contained in the plant biomass [50]. The main parameters that affected the extraction yield of phenolics, while preserving their biological properties, are the condition of the peels, the temperature of the extraction, pH, the types of enzymes, the enzyme concentration, and the citrus species [51]. EAE offers several advantages in pectin extraction from plant resources such as CPW, because this technique can increase the pectin extraction yield compared to traditional processes and can be used under low process temperatures, thereby reducing the energy consumption [52]. EAE also acts as an effective CPW pretreatment method to improve the distillation of EOs, with an increase in extraction yields up to six-fold compared to traditional steam distillation, as well as presents an opportunity for the complete reuse of plant biomass as it generates high amounts of sugars following enzymatic hydrolysis [53]. Finally, EAE can be coupled with UAE, favoring the extraction of compounds most closely bound to the raw material, replacing heat pre-treatment, reducing the ultrasonic extraction time, and consequently increasing the yields by as much as two times [54].

(e) *Subcritical Water Extraction*

The subcritical water extraction (SWE) method involves the use of water at temperature and pressure conditions below the critical point of 374.15 °C and 22.1 MPa. The change in chemical and physical properties allows the selective extraction of polar and nonpolar compounds [55], and the extraction of hydrophilic and lipophilic products [56]. SWE was applied to mandarin (*Citrus unshiu* Markovich) peels at both laboratory (1 g) and pilot (100 g) scales with 10–15 min processing. Total phenolic yields were similar across conditions; hesperidin and narirutin peaked at 130 °C and declined at 150 °C, whereas naringenin was highest at 170 °C [57].

Compared to conventional organic solvent extraction, SWE showed higher extraction yield for hesperidin (1.4–5.8 times), narirutin (1.1–5.6 times), and polymethoxyflavones (from 1.1 to 1.6 times), with the extraction yield of citrus flavonoids directly correlating to temperature and flow rate [58]. These results are consistent with other studies concerning the extraction yield and antioxidant activity of flavanones derived from orange defatted peels [59]. SWE was also used in two-step processes for the extraction of citrus flavanones (hesperidin and narirutin) at 150 °C and mono- and disaccharides at 200 °C [60].

(f) *Natural Deep Eutectic Solvents*

Natural deep eutectic solvents (NADES) are eutectic solvents that use only primary metabolites as solvents, such as sugars, organic acids and bases, and amino acids [35]. NADES can also work in a multistep integrated process to extract different compounds (D-limonene, proteins, and polyphenols) with a unique solvent like cholinium chloride and ethylene glycol as a hydrogen bond donor [61]. The effect of NADES can be enhanced through their combination with UAE, achieving a synergy between mass exchange and disruption of cell membranes performed by UAE, and the stabilization of bioactive compounds provided by NADES [62]. Among the most promising NADES for application to CPW are lactic acid:glucose with a yield for the total phenolic content of 1932 ± 7.83 mgGAE/100 gdw, and L-proline:malic acid with a yield of 2164 ± 5.17 mgGAE/100 gdw, showing also a good stability of polyphenols after 30 days of storage (25 °C and 4 °C) and lower degradability compared to ethanol extracts [63].

(g) *Hydrodynamic Cavitation*

Hydrodynamic cavitation (HC), sharing with UAE the exploitation of the cavitation phenomena in liquid media, is performed either by circulating a liquid through static constrictions of various geometries, or by special immersed rotary equipment.

Contrary to the UAE, HC is a straightforwardly scalable technological solution, showing outstanding effectiveness and efficiency for food processing, process intensification, and extraction of natural products, besides plenty of other applications [64, 65]. Direct pilot-scale experiments, using water as the only solvent, with 42 kg of fresh orange by-products in 120 L of water, the process afforded the extraction of a low methoxyl pectin (degree of esterification of 17.05%) rich in adsorbed hesperidin, naringin, other polyphenols, and terpenes. The application of HC to the extraction of CPW allowed to obtain a completely new class of phytocomplexes, dubbed IntegroPectin [2, 66], with remarkable pharmacological activities, and much higher bioavailability compared to isolated flavonoids. Notably, the role of cavitation in the production of IntegroPectin was later confirmed by experiments using UAE [67].

HC-based extraction processes applied to CPW from lemon and grapefruit also revealed the stable conjugation of pectin with flavonoids and terpenes [68, 69], spontaneously achieving a result comparable to complex manufacturing processes, for example, used to create hydroxytyrosol–pectin conjugates [70]. A HC-based grapefruit extract was used in an in vivo study, showing anti-ischemic cardioprotective activity far exceeding that of the pure bioactive flavanone naringenin on a dose-dependent basis [71].

Finally, it is notable that insoluble residues of HC-based CPW extraction processing mainly consisted of highly micronized cellulose with high technical value [72], thus further contributing to the citrus circular economy.

4 Notes

Table 2 summarizes the main advantages and drawbacks of the considered methods for the extraction of bioactive compounds from CPW.

Across the CPW extraction methods, HC appears to be the most compliant with the principles of green extraction of natural products, including the use of water as the only solvent, efficiency, high-value end products, potential circularity, and scalability [5]. However, the persistent lack of technological and process standardization represents a challenging task that needs to be urgently addressed [73], along with the possible efficiency gains derived from coupling with pretreatment methods such as PEF or EAE.

However, when translated into the production environment, CPW extraction processes are only a step within a complex process chain. As with other bioactive-rich botanicals, formulating end products as water-soluble dry extracts is often necessary for nutraceutical [71], pharmaceutical and cosmeceutical applications [74] to meet requirements for stability, storage, transportation, packaging, and consumer dosing.

The general scheme of the production steps of dry extracts elaborated in a previous study, concerning forestry by-products and using HC in the extraction step [75], also holds for CPW and is shown in Fig. 2.

The main assumptions about the energy balance in the process steps shown in Fig. 2 also hold for the case of CPW.

Figure 3 shows a plausible scenario of the overall and the specific (per unit mass of dry extract) energy consumption, the latter as a measure of process yield, in the process chain leading to the CPW dry extract, assuming a fixed water volume of 1000 L and the use of HC in the extraction step. Quantities are represented as a function of the water to biomass (CPW) ratio (dry weight), with reasonable assumptions for the moisture level of the raw biomass (75%), the extractable yield (30%), and the temperature range of HC processes (20–45 °C), as per the data shown for CPW extraction trials performed with a 200 L pilot installation [2, 71].

Spray drying, followed by vacuum drying, accounts for most of the energy consumption, with the HC-based extraction step accounting for less than 10%. Apparently, the specific energy (energy consumption per unit mass of dry extract) decreases by about 5.5 times with the water to dry biomass ratio decreasing from 20:1 to 1:1, thus, for the sake of economic sustainability, it is advisable to use as high a CPW concentration as possible, which should represent the main goal for fundamental and industrial research. This task is challenging because it involves both practical issues, such as possible clogging, and more substantial issues

Table 2
Main advantages and drawbacks of CPW extraction methods

Method	Advantages	Drawbacks
Conventional methods		
Maceration	Simple and standardized equipment Straightforward scalability Low energy consumption	Further solvent other than water required Long process time process Low extraction yield
Soxhlet	Simple and standardized equipment Straightforward scalability Low energy consumption Good extraction yield Continuous flow extraction process	Critical choice of solvent other than water High volumes of solvent Solvent residues in the end product Long process time Thermal degradation of heat-labile compounds
Hydrodistillation	Simple and standardized equipment Cheap and nontoxic process	Selectivity of EOs extraction Low extraction yield Long process time that could lead to EOs degradation and oxidation of certain compounds
Novel green extraction methods		
Ultrasound-assisted extraction	Water can be the only solvent Low working temperature, fast, and low energy consumption	Non scalable Preservation of bioactive compounds sensitive to working temperature, amplitude, frequency, and power
Pulsed Electric Field	Water as the only solvent Very short processing time	Generally a pre-treatment, needs further extraction technique downstream
Microwave Assisted Extraction	Low working temperature Low energy consumption High extraction yield Preservation of bioactive compounds, including volatiles	Scalability not proven Generally a pre-treatment, needs further extraction technique downstream High cost of equipment at the real scale
Enzyme Assisted Extraction	High quality of recovered pectin As a pretreatment, it allows the UAE to increase the extraction yield of phenolic compounds	Lower recovery of phenolic compounds compared with Soxhlet Selectivity of enzymes Long process time Difficult to scale up High cost of enzymes at the real scale
Subcritical Water Extraction	Water as the only solvent Selective extraction; continuous flow of operation; short time and efficiency	Difficult cleaning Possible degradation of bioactive compounds due to high temperature and pressure High cost of equipment.

(continued)

Table 2 (continued)

Method	Advantages	Drawbacks
Natural Deep Eutectic Solvents	High selectivity of extracted bioactive compounds Low working temperature Simple equipment	Scalability not proven High cost of NADES NADES residues in the end product
Hydrodynamic Cavitation	Water as the only solvent Low working temperature, fast, low energy consumption Creation of new stably conjugated, water-soluble phytocomplexes with higher bioavailability compared to individual compounds Insoluble residues with high technical value Straightforwardly scalable	Nonstandard equipment Critical dependence of performance on construction details, hence the need for new skills

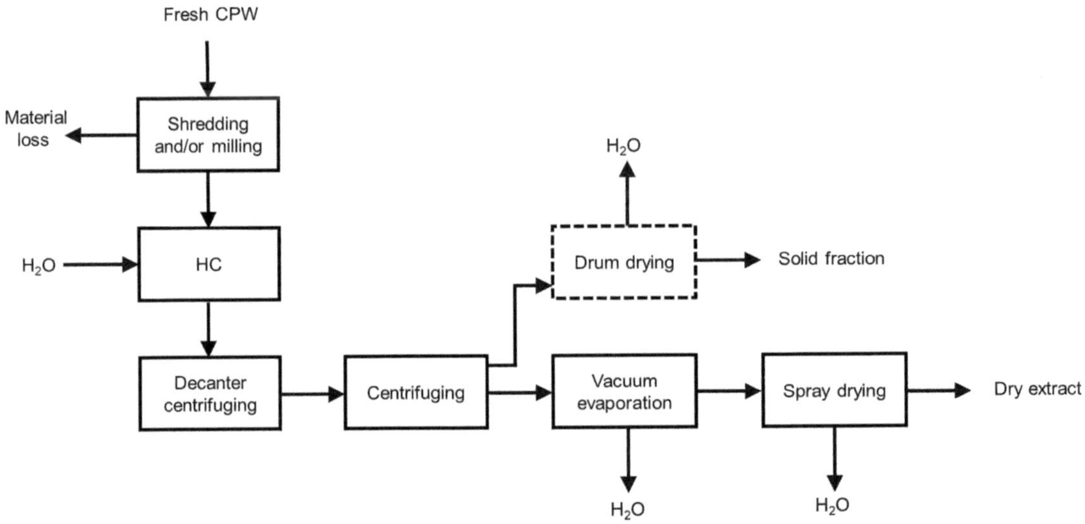

Fig. 2 General scheme of the steps involved in the production of dry extracts using HC in the extraction step. Dashed steps are optional and not considered in the quantitative assessment

deriving from the decreasing cavitation intensity with increasing viscosity of the circulating mixture [76]. The issue of maintaining or intensifying cavitation regimes can be addressed either by means of special structural arrangements of the Venturi reactor [77], or adjusting the cavitation regimes toward optimal working points for cavitation intensity [36, 78].

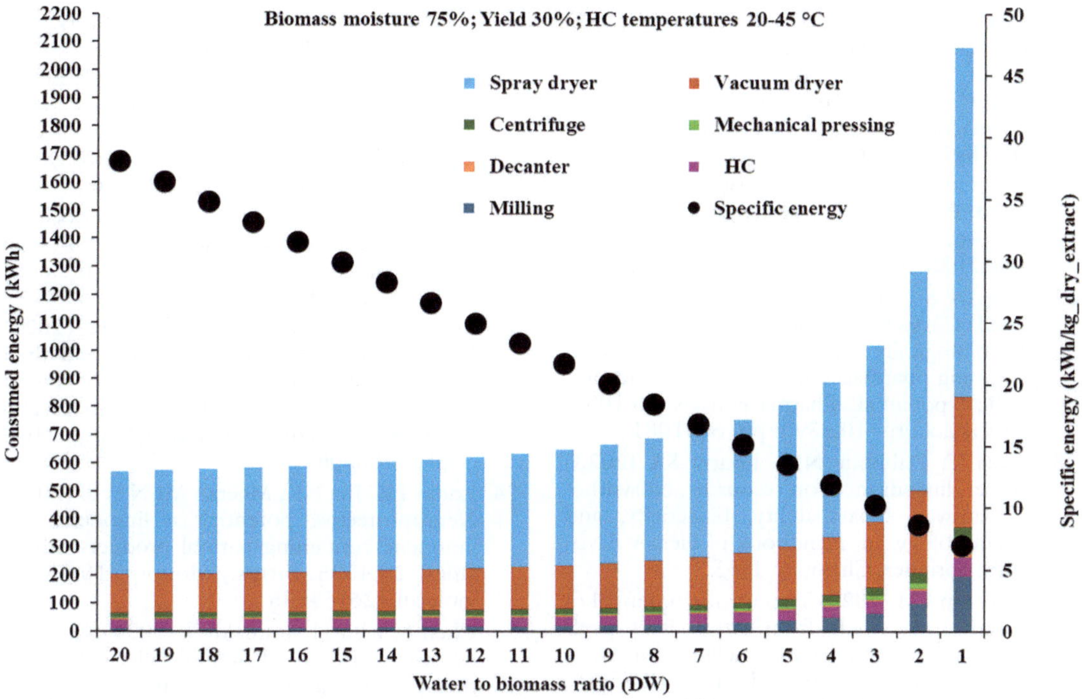

Fig. 3 Scenario of the total energy consumption for a water volume of 1000 L and energy consumption per unit mass of dry extract, as a function of the water to dry biomass ratio, for the chain of process steps leading to the production of dry extracts from raw CPW

References

1. Andrade MA, Barbosa CH, Shah MA et al (2023) Citrus by-products: valuable source of bioactive compounds for food applications. Antioxidants 12:38. https://doi.org/10.3390/antiox12010038
2. Meneguzzo F, Brunetti C, Fidalgo A et al (2019) Real-scale integral valorization of waste orange peel via hydrodynamic cavitation. Processes 7:581. https://doi.org/10.3390/pr7090581
3. Yusuf H, Fors H, Galal NM et al (2025) Barriers to implementing circular citrus supply chains: a systematic literature review. J Environ Manag 373:123963. https://doi.org/10.1016/j.jenvman.2024.123963
4. Meneguzzo F, Zabini F (2021) Agri-food and forestry sectors for sustainable development: innovations to address the ecosystems-resources-climate-food-health nexus. Springer International Publishing, Cham
5. de Souza Mesquita LM, Contieri LS, e Silva FA et al (2024) Path2Green: introducing 12 green extraction principles and a novel metric for assessing sustainability in biomass valorization. Green Chem 26:10087–10106. https://doi.org/10.1039/d4gc02512a
6. (2024) Novel food. In: European Food Safety Authorithy (EFSA). https://www.efsa.europa.eu/en/topics/topic/novel-food
7. Fatimah Wan Mohd Nowalid W, Abd Hamid H, Chandra Iwansyah A et al (2026) Bioactive compounds and antioxidant activity in various citrus peels: a significant systematic review. J Adv Res Appl Sci Eng Technol 55:94–104. https://doi.org/10.37934/araset.55.1.94104
8. Sharma P, Vishvakarma R, Gautam K et al (2022) Valorization of citrus peel waste for the sustainable production of value-added products. Bioresour Technol 351:127064. https://doi.org/10.1016/j.biortech.2022.127064
9. Chandel V, Biswas D, Roy S et al (2022) Current advancements in pectin: extraction, properties and multifunctional applications. Foods 11:2683. https://doi.org/10.3390/FOODS11172683

10. Maslova AY, Mishvelov AE, Nasrulaeva KN et al (2023) Overview of the pharmacological use of pectins and pectin-containing substances: recent achievements and prospects. Pharmacophore 14:39–44. https://doi.org/10.51847/j34k56lsvr
11. Zaitseva O, Khudyakov A, Sergushkina M et al (2020) Pectins as a universal medicine. Fitoterapia 146:104676. https://doi.org/10.1016/j.fitote.2020.104676
12. Šafranko S, Šubarić D, Jerković I, Jokić S (2023) Citrus by-products as a valuable source of biologically active compounds with promising pharmaceutical, biological and biomedical potential. Pharmaceuticals 16:1081. https://doi.org/10.3390/ph16081081
13. Donia T, Dabbour NM, Loutfy SA (2023) Hesperidin: advances on resources, biosynthesis pathway, bioavailability, bioactivity, and pharmacology. In: Handbook of dietary flavonoids. Springer, Cham, pp 1–55
14. Lu K, Yip YM (2023) Therapeutic potential of bioactive flavonoids from citrus fruit peels toward obesity and diabetes mellitus. Fut Pharmacol 3:14–37. https://doi.org/10.3390/FUTUREPHARMACOL3010002
15. Meneguzzo F, Ciriminna R, Zabini F, Pagliaro M (2020) Review of evidence available on hesperidin-rich products as potential tools against COVID-19 and hydrodynamic cavitation-based extraction as a method of increasing their production. Processes 8:549. https://doi.org/10.3390/PR8050549
16. Rahmani AH, Babiker AY, Anwar S (2023) Hesperidin, a bioflavonoid in cancer therapy: a review for a mechanism of action through the modulation of cell signaling pathways. Molecules 28:5152. https://doi.org/10.3390/molecules28135152
17. Rodrigues CV, Pintado M (2024) Hesperidin from orange peel as a promising skincare bioactive: an overview. Int J Mol Sci 25:1890. https://doi.org/10.3390/ijms25031890
18. Bellavite P (2023) Neuroprotective potentials of flavonoids: experimental studies and mechanisms of action. Antioxidants 12:280. https://doi.org/10.3390/ANTIOX12020280
19. Testai L, Da Pozzo E, Piano I et al (2017) The citrus flavanone naringenin produces cardioprotective effects in hearts from 1 year old rat, through activation of mitoBK channels. Front Pharmacol 8. https://doi.org/10.3389/fphar.2017.00071
20. Naeini F, Namkhah Z, Tutunchi H et al (2022) Effects of naringenin supplementation on cardiovascular risk factors in overweight/obese patients with nonalcoholic fatty liver disease: a pilot double-blind, placebo-controlled, randomized clinical trial. Eur J Gastroenterol Hepatol 34:345–353. https://doi.org/10.1097/MEG.0000000000002323
21. Adetunji JA, Fasae KD, Awe AI et al (2023) The protective roles of citrus flavonoids, naringenin, and naringin on endothelial cell dysfunction in diseases. Heliyon 9. https://doi.org/10.1016/j.heliyon.2023.e17166
22. Yao L, Liu W, Bashir M et al (2022) Eriocitrin: a review of pharmacological effects. Biomed Pharmacother 154:113563. https://doi.org/10.1016/j.biopha.2022.113563
23. Antonelli M, Donelli D, Barbieri G et al (2020) Forest volatile organic compounds and their effects on human health: a state-of-the-art review. Int J Environ Res Public Health 17:1–36. https://doi.org/10.3390/ijerph17186506
24. Eddin LB, Jha NK, Meeran MFN et al (2021) Neuroprotective potential of limonene and limonene containing natural products. Molecules 26:4535. https://doi.org/10.3390/molecules26154535
25. Chen S, Zhu H, Jounaidi Y (2024) Comprehensive snapshots of natural killer cells functions, signaling, molecular mechanisms and clinical utilization. Signal Transduct Target Ther 9:302. https://doi.org/10.1038/s41392-024-02005-w
26. Wu J, Wang Q, Xu C et al (2024) Biogenic volatile organic compounds in forest therapy base: a source of air pollutants or a healthcare function? Sci Total Environ 931: 172944. https://doi.org/10.1016/j.scitotenv.2024.172944
27. Moreira MM, Morais S, Delerue-Matos C (2017) Environment-friendly techniques for extraction of bioactive compounds from fruits. In: Soft chemistry and food fermentation. Elsevier, pp 21–47. https://doi.org/10.1016/B978-0-12-811412-4.00002-3
28. Hussain A, Gulbadan Dar N, Paracha GM, Akhter S (2015) Evaluation of different techniques for extraction of antioxidants as bioactive compounds from citrus peels (industrial by products). J Agric Environ Sci 15:676–682. https://doi.org/10.5829/idosi.aejaes.2015.15.4.12604
29. Safdar MN, Kausar T, Jabbar S et al (2017) Extraction and quantification of polyphenols from kinnow (Citrus reticulate L.) peel using ultrasound and maceration techniques. J Food Drug Anal 25:488–500. https://doi.org/10.1016/j.jfda.2016.07.010
30. Al-Juhaimi FY (2014) Citrus fruits by-products as sources of bioactive compounds with antioxidant potential. Pak J Bot 46(4):1459–1462
31. Shofinita D, Feng S, Langrish TAG (2015) Comparing yields from the extraction of

different citrus peels and spray drying of the extracts. Adv Powder Technol 26:1633–1638. https://doi.org/10.1016/j.apt.2015.09.007
32. Jagannath A, Biradar R (2019) Comparative evaluation of soxhlet and ultrasonics on the structural morphology and extraction of bioactive compounds of lemon (Citrus limon l.) peel. J Food Chem Nanotechnol 5:56–64. https://doi.org/10.17756/jfcn.2019-072
33. Gilani F, Raftani Amiri Z, Esmaeilzadeh Kenari R, Ghaffari Khaligh N (2023) Investigation of extraction yield, chemical composition, bioactive compounds, antioxidant and antimicrobial characteristics of citron (Citrus medica L.) peel essential oils produced by hydrodistillation and supercritical carbon dioxide. J Food Meas Charact 17:4332–4344. https://doi.org/10.1007/s11694-023-01956-y
34. Shaw D, Tripathi AD, Paul V et al (2023) Valorization of essential oils from citrus peel powder using hydro-distillation. Sustain Chem Pharm 32. https://doi.org/10.1016/j.scp.2023.101036
35. Fernández-Cabal J, Avilés-Betanzos KA, Cauich-Rodríguez JV et al (2025) Recent developments in Citrus aurantium L.: an overview of bioactive compounds, extraction techniques, and technological applications. Processes 13:120. https://doi.org/10.3390/pr13010120
36. Ge M, Zhang G, Petkovšek M et al (2022) Intensity and regimes changing of hydrodynamic cavitation considering temperature effects. J Clean Prod 338:130470. https://doi.org/10.1016/j.jclepro.2022.130470
37. Acciardo E, Tabasso S, Cravotto G, Bensaid S (2022) Process intensification strategies for lignin valorization. Chem Eng Process Process Intensif 171:108732. https://doi.org/10.1016/j.cep.2021.108732
38. Chemat F, Rombaut N, Sicaire A-G et al (2017) Ultrasound assisted extraction of food and natural products. Mechanisms, techniques, combinations, protocols and applications. A review. Ultrason Sonochem 34:540–560. https://doi.org/10.1016/j.ultsonch.2016.06.035
39. Papoutsis K, Pristijono P, Golding JB et al (2018) Screening the effect of four ultrasound-assisted extraction parameters on hesperidin and phenolic acid content of aqueous citrus pomace extracts. Food Biosci 21:20–26. https://doi.org/10.1016/j.fbio.2017.11.001
40. Vathsala V, Singh SP, Bishnoi M et al (2024) Ultrasound-assisted extraction (UAE) and characterization of citrus peel pectin: comparison between pummelo (Citrus grandis L. Osbeck) and sweet lime (Citrus limetta Risso). Sustain Chem Pharm 37. https://doi.org/10.1016/j.scp.2023.101357
41. Kaur S, Panesar PS, Chopra HK (2023) Citrus processing by-products: an overlooked repository of bioactive compounds. Crit Rev Food Sci Nutr 63:67–86. https://doi.org/10.1080/10408398.2021.1943647
42. Peiró S, Luengo E, Segovia F et al (2019) Improving polyphenol extraction from lemon residues by pulsed electric fields. Waste Biomass Valorization 10:889–897. https://doi.org/10.1007/s12649-017-0116-6
43. Chatzimitakos T, Athanasiadis V, Kalompatsios D et al (2023) Pulsed electric field applications for the extraction of bioactive compounds from food waste and by-products: a critical review. Biomass (Switzerland) 3:367–401
44. Hwang HJ, Kim HJ, Ko MJ, Chung MS (2021) Recovery of hesperidin and narirutin from waste Citrus unshiu peel using subcritical water extraction aided by pulsed electric field treatment. Food Sci Biotechnol 30:217–226. https://doi.org/10.1007/s10068-020-00862-z
45. Luengo E, Álvarez I, Raso J (2013) Improving the pressing extraction of polyphenols of orange peel by pulsed electric fields. Innov Food Sci Emerg Technol 17:79–84. https://doi.org/10.1016/j.ifset.2012.10.005
46. Carpentieri S, Režek Jambrak A, Ferrari G, Pataro G (2022) Pulsed electric field-assisted extraction of aroma and bioactive compounds from aromatic plants and food by-products. Front Nutr 8. https://doi.org/10.3389/fnut.2021.792203
47. García-Martín JF, Feng CH, Domínguez-Fernández NM, Álvarez-Mateos P (2023) Microwave-assisted extraction of polyphenols from bitter orange industrial waste and identification of the main compounds. Life 13. https://doi.org/10.3390/life13091864
48. Nayak B, Dahmoune F, Moussi K et al (2015) Comparison of microwave, ultrasound and accelerated-assisted solvent extraction for recovery of polyphenols from Citrus sinensis peels. Food Chem 187:507–516. https://doi.org/10.1016/j.foodchem.2015.04.081
49. Alvi T, Asif Z, Iqbal Khan MK (2022) Clean label extraction of bioactive compounds from food waste through microwave-assisted extraction technique-a review. Food Biosci 46. https://doi.org/10.1016/j.fbio.2022.101580
50. Patra A, Abdullah S, Pradhan RC (2022) Review on the extraction of bioactive

compounds and characterization of fruit industry by-products. Bioresour Bioprocess 9. https://doi.org/10.1186/s40643-022-00498-3

51. Li BB, Smith B, Hossain MM (2006) Extraction of phenolics from citrus peels: II. Enzyme-assisted extraction method. Sep Purif Technol 48:189–196. https://doi.org/10.1016/j.seppur.2005.07.019
52. Marić M, Grassino AN, Zhu Z et al (2018) An overview of the traditional and innovative approaches for pectin extraction from plant food wastes and by-products: ultrasound-, microwaves-, and enzyme-assisted extraction. Trends Food Sci Technol 76:28–37
53. Chávez-González ML, López-López LI, Rodríguez-Herrera R et al (2016) Enzyme-assisted extraction of citrus essential oil. Chem Pap 70:412–417. https://doi.org/10.1515/chempap-2015-0234
54. Durmus N, Kilic-Akyilmaz M (2023) Bioactivity of non-extractable phenolics from lemon peel obtained by enzyme and ultrasound assisted extractions. Food Biosci 53. https://doi.org/10.1016/j.fbio.2023.102571
55. Brezo-borjan T, Švarc-gaji J, Rodrigues F (2023) Chemical and biological characterisation of orange. Processes 11:1766. https://doi.org/10.3390/pr11061766
56. Costa JM, Strieder MM, Saldaña MDA et al (2023) Recent advances in the processing of agri-food by-products by subcritical water. Food Bioprocess Technol 16:2705–2724
57. Ko MJ, Kwon HL, Chung MS (2016) Pilot-scale subcritical water extraction of flavonoids from satsuma mandarin (Citrus unshiu Markovich) peel. Innov Food Sci Emerg Technol 38: 175–181. https://doi.org/10.1016/j.ifset.2016.10.008
58. Kim DS, Bin LS (2020) Kinetic study of subcritical water extraction of flavonoids from citrus unshiu peel. Sep Purif Technol 250. https://doi.org/10.1016/j.seppur.2020.117259
59. Lachos-Perez D, Baseggio AM, Mayanga-Torres PC et al (2018) Subcritical water extraction of flavanones from defatted orange peel. J Supercrit Fluids 138:7–16. https://doi.org/10.1016/j.supflu.2018.03.015
60. Lachos-Perez D, Baseggio AM, Torres-Mayanga PC et al (2020) Sequential subcritical water process applied to orange peel for the recovery flavanones and sugars. J Supercrit Fluids 160. https://doi.org/10.1016/j.supflu.2020.104789
61. Panić M, Andlar M, Tišma M et al (2021) Natural deep eutectic solvent as a unique solvent for valorisation of orange peel waste by the integrated biorefinery approach. Waste Manag 120:340–350. https://doi.org/10.1016/j.wasman.2020.11.052
62. Mayanin I (2024) Evaluation of polyphenol profile from citrus peel obtained by natural deep eutectic solvent / ultrasound extraction. Processes 12:1–19. https://doi.org/10.3390/pr12102072
63. Gómez-Urios C, Viñas-Ospino A, Puchades-Colera P et al (2022) Sustainable development and storage stability of orange by-products extract using natural deep eutectic solvents. Foods 11. https://doi.org/10.3390/foods11162457
64. Ciriminna R, Scurria A, Pagliaro M (2023) Natural product extraction via hydrodynamic cavitation. Sustain Chem Pharm 33:101083. https://doi.org/10.1016/j.scp.2023.101083
65. Arya SS, More PR, Ladole MR et al (2023) Non-thermal, energy efficient hydrodynamic cavitation for food processing, process intensification and extraction of natural bioactives: a review. Ultrason Sonochem 98:106504. https://doi.org/10.1016/j.ultsonch.2023.106504
66. Scurria A, Sciortino M, Garcia AR et al (2022) Red orange and bitter orange IntegroPectin: structure and main functional compounds. Molecules 27. https://doi.org/10.3390/molecules27103243
67. Ciriminna R, Angellotti G, Li Petri G et al (2024) Cavitation as a zero-waste circular economy process to convert citrus processing waste into biopolymers in high demand. J Bioresour Bioprod 9:486–494. https://doi.org/10.1016/j.jobab.2024.09.002
68. Scurria A, Sciortino M, Albanese L et al (2021) Flavonoids in lemon and grapefruit IntegroPectin. ChemistryOpen 10:1055–1058. https://doi.org/10.1002/open.202100223
69. Scurria A, Sciortino M, Presentato A et al (2021) Volatile compounds of lemon and grapefruit IntegroPectin. Molecules 26:51. https://doi.org/10.3390/molecules26010051
70. Rubio-Senent F, Bermúdez-Oria A, Rodríguez-Gutiérrez G et al (2025) Structural and antioxidant properties of hydroxytyrosol-pectin conjugates: comparative analysis of adsorption and free radical methods and their impact on in vitro gastrointestinal process. Food Hydrocoll 162:110954. https://doi.org/10.1016/j.foodhyd.2024.110954
71. Flori L, Albanese L, Calderone V et al (2022) Cardioprotective effects of grapefruit IntegroPectin extracted via hydrodynamic cavitation from by-products of citrus fruits industry: role

of mitochondrial potassium channels. Foods 11:2799. https://doi.org/10.3390/FOODS11182799

72. Al Jitan S, Scurria A, Albanese L et al (2022) Micronized cellulose from citrus processing waste using water and electricity only. Int J Biol Macromol 204:587–592. https://doi.org/10.1016/j.ijbiomac.2022.02.042
73. Iyer G, Pandit AB (2024) Bridging ingenuity and utility in cavitation—a pioneer's predicament. Ind Eng Chem Res 63:12265–12276. https://doi.org/10.1021/acs.iecr.4c01817
74. Sebastian Raj A (2023) Advancing phytonutrient extraction via cavitation-based methodology: exploring catechin recovery from Camellia sinensis leaves. Biocatal Agric Biotechnol 54: 102895. https://doi.org/10.1016/j.bcab.2023.102895
75. Tienaho J, Liimatainen J, Myllymäki L et al (2024) Pilot scale hydrodynamic cavitation and hot-water extraction of Norway spruce bark yield antimicrobial and polyphenol-rich fractions. Sep Purif Technol 360:130925. https://doi.org/10.1016/j.seppur.2024.130925
76. Luo J, Fu G, Xu W et al (2024) Experimental study on attenuation effect of liquid viscosity on shockwaves of cavitation bubbles collapse. Ultrason Sonochem 111:107063. https://doi.org/10.1016/j.ultsonch.2024.107063
77. Ding W, Hong F, Ying D et al (2024) A comprehensive study on the effects of annular protrusion for cavitation intensification in Venturi tubes. Chem Eng J 498:155306. https://doi.org/10.1016/j.cej.2024.155306
78. Albanese L, Ciriminna R, Meneguzzo F, Pagliaro M (2015) Energy efficient inactivation of Saccharomyces cerevisiae via controlled hydrodynamic cavitation. Energy Sci Eng 3: 221–238. https://doi.org/10.1002/ese3.62

Chapter 3

Total Phenols in Citrus By-Products

Adriana Gadioli Tarone and Cinthia Baú Betim Cazarin

Abstract

The quantification of phenolic compounds in foods has become increasingly important in recent years because of the numerous health benefits associated with their consumption. Citrus fruits, particularly their by-products, contain various phenolic compounds that can be reused and their bioactive components can be extracted. This chapter discusses a widely used methodology from the literature for evaluating the total content of polyphenols, highlighting its advantages and limitations.

Key words Phenolic compounds, Polyphenols, Bioactive compounds, Citrus by-products, Antioxidant activity

1 Introduction

Citrus fruits are among the most widely cultivated fruits in the world, and citrus juices represent a significant segment of the international beverage market. In the processing of citrus juice, more than 50% of the fresh fruit's weight is transformed into organic by-products. When these by-products are discarded without proper treatment, they can cause considerable harm to natural resources and the environment. Therefore, it is essential to manage this substantial volume of by-products to ensure an environmentally sustainable processing chain [1, 2].

Citrus by-products include the peel, seeds, and internal fruit components (mesocarp and endocarp), all of which are rich in various bioactive compounds, including polyphenols. Regular consumption of these compounds has been consistently associated with health benefits, offering advantages such as antiviral, antioxidant, antifungal, anti-inflammatory, and anticarcinogenic activities, thus demonstrating significant nutraceutical potential [3–5]. Given the high phenolic content in citrus by-products, it is crucial to quantify these compounds quickly and accurately [6].

Shweta Suri et al. (eds.), *Citrus Fruit Processing By-Products: Technologies and Basic Protocols*, Methods and Protocols in Food Science, https://doi.org/10.1007/978-1-0716-5088-2_3,

Phenolic compounds are secondary metabolites found abundantly in fruits and vegetables, and can be classified into several categories based on their chemical structure [7, 8]. The method used to determine the total phenolic content in citrus by-products involves oxidizing agents that react with phenolic compounds to produce colored compounds. The oxidizing agent commonly used is the Folin–Ciocalteu reagent, which reacts with phenols in the presence of sodium carbonate to form a blue-colored complex that absorbs light at 725 nm. The intensity of this blue color is directly proportional to the concentration of reactive phenolic compounds present in the analyzed sample [9–12].

While the Folin–Ciocalteu reagent is commonly used to quantify total phenolics, it is crucial to note that the Folin–Ciocalteu reagent lacks specificity for the phenol group. As a result, it can also react with other substances, such as vitamins, sugars, and amino acids. Therefore, this methodology should evaluate total reducing capacity or total antioxidant capacity rather than total phenolics [10].

High-performance liquid chromatography (HPLC) is the best method to accurately quantify phenolics, although it is a more expensive technique. This technique can be coupled with mass spectrometry (MS) for enhanced analysis. In addition to having the necessary analytical equipment, you will need analytical standards or access to a database to identify fragments through MS.

Another vital aspect of quantifying phenolics is sample preparation. Various extraction techniques and solvents typically extract compounds from the food matrix. The choice of solvent directly impacts the type of compounds extracted; nonpolar compounds will be extracted with nonpolar solvents, while polar compounds will require polar solvents. Depending on the extraction technique used, only a fraction of the compounds may be extracted, or a complete extraction (exhaustive) might occur. It is important to consider the type of extract you are working with, as the results may reflect the phenolic content of the extract rather than the entire food matrix.

Furthermore, the chemical properties of the extracted compounds can influence their extraction and storage conditions. These characteristics can lead to degradation and an inaccurate estimation of their content in the food. Some phenolic compounds may also be strongly bonded to the food matrix structures, necessitating specific methodologies, like enzymes, to facilitate their extraction [13].

Considering all the above points, this chapter will outline the colorimetric method for quantifying reducing capacity, commonly called total phenolics quantification.

2 Materials

All solutions must be prepared using ultrapure water (obtained by purifying deionized water, with 18 MΩ-cm at 25 °C of sensibility) and analytical-grade reagents. When handling and disposing of waste materials, ensure strict adherence to all waste disposal regulations.

1. Folin–Ciocalteu's phenol ready-made reagent (Cat. N° F9252).
2. Sodium Carbonate Solution: 10% sodium carbonate in water (w/v). Weigh 1 g of sodium carbonate (Cat. N° 7527) in a 10 mL beaker and add 8 mL of water. Dilute the solution using a bar magnet and a magnetic stirrer. Transfer the solution to a 10 mL volumetric flask and add water until the meniscus aligns with the calibration mark (*see* **Notes 1–3**).
3. A *UV–Vis* spectrophotometer microplate reader for the colorimetric analysis (*see* **Note 4**).
4. 96-well microplates.
5. Micropipettes (1–10 mL) (*see* **Note 5**).
6. Eppendorf tubes (1.5–2 mL).
7. Vortex mixer.
8. Bar magnet and magnetic stirrer.
9. Ultrasound bath.

2.1 Preparation of Calibration Standard Solutions

1. Gallic acid standard stock solution 1 (1 mg/L). Weigh 0.01 g of gallic acid (Cat. N° G7384) in a 10 mL beaker and add 8 mL of water. Dilute the solution using a bar magnet and a magnetic stirrer. Transfer the solution to a 10 mL volumetric flask and add water until the meniscus aligns with the calibration mark (*see* **Notes 6** and 7).
2. Prepare calibration standard solutions as described in Table 1 (16–500 μg/L of gallic acid) by pipetting the specified volume of stock standard solution into a 1.5 mL Eppendorf tube and diluting it with the specified volume of water. Mix well using a vortex mix for approximately 30 s. Table 1 presents the approximate concentrations of gallic acid in each calibration standard solution. We suggest this concentration range, but you can change this range (*see* **Note 8**).
3. The solutions must be prepared and analyzed in triplicate for greater accuracy of the calibration standard curve.

Table 1
Preparation of calibration standard solutions

Calibration standard solution	Standard stock solution 2 volume (mL)	Water volume (mL)	Gallic acid approximate concentration (μg/L)
1	16	984	16
2	25	975	25
3	50	950	50
4	75	925	75
5	100	900	100
6	250	750	250
7	500	500	500

2.2 Preparation of Sample Solutions

1. Prepare your sample based on the characteristics of the food matrix to be evaluated.
2. If you will analyze fruit juice, you can only filter and dilute the sample.
3. The solutions must be prepared and analyzed in triplicate for greater accuracy of the sample results.

3 Methods

3.1 Total Phenolic Compounds Analysis

1. Prepare 1.5 mL Eppendorf tubes for each standard curve concentration, for the samples and blank.
2. Create a template that indicates each well on the plate and its corresponding sample, as illustrated in Fig. 1.
3. Add 800 μL of water and 50 μL of Folin–Ciocalteu's phenol ready-made reagent to a 1.5 mL Eppendorf tube.
4. To each Eppendorf tube, add one of the following: (1) 50 μL of the sample solution, (2) 50 μL of calibration standard solutions 1–6, or (3) 50 μL of water (blank). Mix the contents of each tube thoroughly using a vortex mixer for approximately 30 s, then let them sit in the dark for 3 min.
5. Add 100 μL of sodium carbonate solution to each Eppendorf tube and mix thoroughly using a vortex mixer for approximately 30 s. Allow the tubes to sit in the dark for 120 min.
6. Transfer 250 μL from each Eppendorf tube after the reaction into a well of a 96-well microplate, following your template.
7. Measure the absorbance at 725 nm using a spectrophotometer, zeroing it with the blank solution.

	1	2	3	4	5	6	7	8	9	10	11	12
A	Blank	Blank	Blank	Sample 2	Sample 2	Sample 2						
B	Std 1	Std 1	Std 1	Sample 3	Sample 3	Sample 3						
C	Std 2	Std 2	Std 2	Sample 4	Sample 4	Sample 4						
D	Std 3	Std 3	Std 3	Sample 5	Sample 5	Sample 5						
E	Std 4	Std 4	Std 4	Sample 6	Sample 6	Sample 6						
F	Std 5	Std 5	Std 5	Sample 7	Sample 7	Sample 7						
G	Std 6	Std 6	Std 6	Sample 8	Sample 8	Sample 8						
H	Sample 1	Sample 1	Sample 1	Sample 9	Sample 9	Sample 9						

Fig. 1 Template of the 96-well microplates

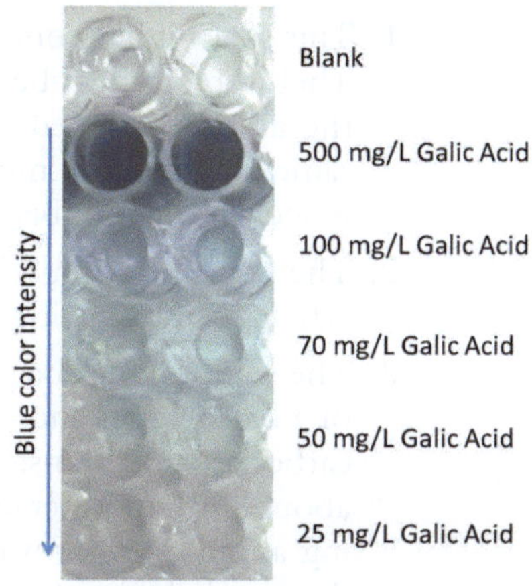

Fig. 2 Example of a microplate prepared to quantify total phenolics. (Created by the authors)

8. Your standard curve should exhibit a blue color and an increase in absorbance as the concentration rises (Fig. 2).
9. The dilution of the sample solution can be adjusted based on the total phenolic content to ensure the absorbance falls within the calibration curve range.

3.2 Calculations

1. Calibration Curve: Plot the absorbance at 725 nm against the concentrations of the calibration standard solutions. Use linear regression (forced through zero) to determine the slope, *y*-intercept, and R^2 value of the calibration curve for gallic acid. Ensure the data and curve fit are visually inspected to confirm consistency with the linear model (*see* **Note 9**). The R^2 value of the calibration curve must be ≥ 0.990 [9].
2. Sample Solutions: Calculate the total phenolic content of the sample solutions using the following formula:

$$y = ax + b$$

where y = the absorbance of the sample solution at 725 nm; b = the *y*-intercept of the calibration curve; a = the slope of the calibration curve. Then, x = mg of total phenolics in gallic acid equivalent / L of sample solution).

3. If dilution of the sample solution is necessary, the total phenolic content calculated must be multiplied by the dilution factor to obtain the final result.

4 Notes

1. This is a highly concentrated salt solution with low stability. Therefore, it must be prepared on the day of use. Prepare only the amount needed and discard any unused solution on the same day. You can modify the solution volume to be prepared based on the number of samples for evaluation.
2. The concentration of sodium carbonate reported in scientific articles can vary widely, from 2% to 35%.
3. The solution may be prepared using an ultrasound bath instead of a bar magnet and magnetic stirrer. Weigh 1.06 g of sodium carbonate and transfer it into a 10 mL volumetric flask. Add about 8 mL of water and sonicate in an ultrasonic bath containing about 3 cm of water for up to 10 min until the solids are dissolved. Add water until the meniscus aligns with the calibration mark and mix well.
4. Using a microplate reader allows you to optimize your time by analyzing more samples per plate and reducing the required reagents. However, if you cannot access a microplate reader, you can adjust the volumes and perform your readings using a cuvette.
5. Make sure your micropipettes are properly calibrated. The small volumes used can impact the accuracy of your analysis.
6. The solution may be prepared using an ultrasound bath instead of a bar magnet and magnetic stirrer. Weigh 0.01 g of gallic acid and transfer it into a 10 mL volumetric flask. Add about

8 mL of water and sonicate in an ultrasonic bath containing about 3 cm of water for up to 10 min until the solids are dissolved. Add water until the meniscus aligns with the calibration mark and mix well.

7. Oxygen and light exposure can easily degrade gallic acid dilute. Therefore, it must be prepared on the day of use. Discard any unused solution on the same day.
8. You can adjust the suggested concentrations to better fit your laboratory and sample conditions. Just be mindful of the dilution calculations.
9. When assessing the linearity of your calibration curve, consider not only the R^2 values but also the reproducibility of your results. Always verify that the absorbance values are within the expected range and check the coefficient of variation between replicates. R^2 values alone should not be used to determine the quality of your standard curve.

Acknowledgments

The authors acknowledge the Coordenação de Aperfeiçoamento de Pessoal de Nível Superior—Brasil (CAPES)—Finance Code 001 and the National Council for Scientific and Technological Development (CNPq) (306891/2021-2).
The authors used generative AI from Grammarly to improve the quality of text.

References

1. Suri S, Singh A, Nema PK (2022) Current applications of citrus fruit processing waste: a scientific outlook. Appl Food Res 2(1):100050
2. Ortiz-Sanchez M, Omarini AB, González-Aguirre JA, Baglioni M, Zygadlo JA, Breccia J et al (2023) Valorization routes of citrus waste in the orange value chain through the biorefinery concept: the Argentina case study. Chem Eng Process Process Intensif 189:109407
3. Gadi S, Pérez-Vega S, Minjares-Fuentes R, Morales-Oyervides L, Contreras-Esquivel JC, Montañez J (2024) Novel extraction technologies for the recovery of bioactive compounds from citrus by-products: recent findings. In: Torre, L.S., Contreras-Esquivel, J.C., Abraham, A.R., & Haghi, A.K. (Eds.). (2024). Bioresources and Bioprocess in Biotechnology for a Sustainable Future (1st ed.). Apple Academic Press, New York. https://doi.org/10.1201/9781003410041
4. Kaur S, Panesar PS, Chopra HK (2021) Citrus processing by-products: an overlooked repository of bioactive compounds. Food Sci Nutr 63(1):67–86
5. Mateus ARS, Pena A, Sanches-Silva A (2024) Unveiling the potential of bioactive compounds in vegetable and fruit by-products: exploring phytochemical properties, health benefits, and industrial opportunities. Curr Opin Green Sustain Chem 48:100938
6. Mateus ARS, Mariño-Cortegoso S, Barros SC, Sendón R, Barbosa L, Pena A et al (2024) Citrus by-products: a dual assessment of antioxidant properties and food contaminants towards circular economy. Innovative Food Sci Emerg Technol 95:103737
7. Lang Y, Gao N, Zang Z, Meng X, Lin Y, Yang S et al (2024) Classification and antioxidant assays of polyphenols: a review. J Future Foods 4(3):193–204

8. Del Rio D, Rodriguez-Mateos A, Spencer JPE, Tognolini M, Borges G, Crozier A (2013) Dietary (poly)phenolics in human health: structures, bioavailability, and evidence of protective effects against chronic diseases. Antioxid Redox Signal 18:1818–1892
9. Kupina S, Fields C, Roman MC, Brunelle SL (2018) Determination of total phenolic content using the Folin-C assay: single-laboratory validation, first action 2017.13. J AOAC Int 102(1):320–321
10. Everette JD, Bryant QM, Green AM, Abbey YA, Wangila GW, Walker RB (2010) Thorough study of reactivity of various compound classes toward the folin-ciocalteu reagent. J Agric Food Chem 58(14):8139–8144
11. Singleton VL, Orthofer R, Lamuela-Raventós RM (1999) [14] Analysis of total phenols and other oxidation substrates and antioxidants by means of folin-ciocalteu reagent. Methods Enzymol 299:152–178
12. Swain T, Hillis WE (1959) The phenolic constituents of Prunus domestica. I.—The quantitative analysis of phenolic constituents. J Sci Food Agric 10(1):63–68
13. Acosta-Estrada BA, Gutiérrez-Uribe JA, Serna-Saldívar SO (2014) Bound phenolics in foods, a review. Food Chem. Elsevier Ltd 152:46–55

Chapter 4

Total Flavonoids in Citrus By-Products

Afiya John and Omina Shomirzoeva

Abstract

Determining the flavonoid content in citrus by-products is essential for evaluating their potential health benefits and optimizing their use in nutraceuticals, functional foods, and pharmaceuticals. Citrus by-products represent a valuable and underexploited source of flavonoids, which can be extracted using various methods such as solvent extraction followed by quantification by colorimetric, HPLC, or spectrophotometric assay. The development of efficient and sustainable methods to extract and quantify flavonoids from citrus by-products not only offers opportunities for the valorization of food waste but also contributes to the growing field of natural antioxidants and nutraceuticals.

Key words Citrus Peels, Flavonoids, HPLC, Extraction

1 Introduction

Citrus has historically been valued both as a food source and as a medicinal plant. Citrus fruits are rich in flavonoids, and their extracts are increasingly being used as functional components for several industrial products [1]. Citrus fruits, their derivatives, and their by-products (peel, pulp, and oil) are rich in bioactive flavonoids such as anthocyanins, flavones, and flavonols; stilbenes, such as trans-resveratrol; and phenolic acids, such as gallic, vanillic, syringic, and caffeic [2]. During the processing of citrus juice, peels are the main by-product and a source of environmental pollution [2].

Citrus flavonoids possess potential antioxidant properties (prevent aging), anticancer, antiviral, and anti-inflammatory activities, effects on capillarity, and cholesterol-lowering ability [3–5]. Citrus fruits are well known for their dietary, nutritional, medicinal, and cosmetic properties and are also good sources of citric acid, flavonoids, phenolics, pectins, limonoids, ascorbic acid, etc. Citrus fruits, including oranges, lemons, limes, grapefruits, etc., are a principal source of essential nutrients, which are suggested to be the main bioactivities responsible for the prevention of

Shweta Suri et al. (eds.), *Citrus Fruit Processing By-Products: Technologies and Basic Protocols*, Methods and Protocols in Food Science, https://doi.org/10.1007/978-1-0716-5088-2_4,

degenerative disease. These include vitamins C, folic acid, carotenoids, dietary fibers, potassium, selenium, and a wide range of phytochemicals [3].

1.1 Flavonoids in Citrus Fruit By-Products

Citrus flavonoids are polyphenolic compounds found specifically in citrus fruits and plants, such as oranges, mandarins, grapefruit, lemons, bergamots, and limes, and they play a crucial role in maintaining physical health. Flavonoids are a group of naturally occurring polyphenolic compounds that have the same basic skeleton of 15 carbons (C_6–C_3–C_6), consisting of two phenyl rings (A and B) linked by a heterocyclic pyran or pyrone ring (C) in the center, depending on their substituents. Flavonoids are further subdivided into flavanols, anthocyanidins, flavanones, flavones, and chalcones. The generic flavonoid structure and the numbering system used to distinguish the carbon positions around the molecule. The three phenolic rings that make up the flavonoid molecule are called pyran rings. Flavanones, flavones, and flavonols are three principal flavonoids in citrus species. The main flavonoids found in citrus species are hesperidin, narirutin, naringin, and eriocitrin [6].

Flavanones

Flavanones are generally the most abundant individual components of the flavonoid pool of citrus fruits and juices. Hesperidin is the most abundant flavanone glycoside found in citrus peel [7]. The highest concentrations of accumulation of flavanones (Fig. 1) are found in the peel as compared to the pulp part of the fruit.

Flavones

Flavones are another group that is abundant in citrus, following flavanones. Overall, in most citrus juices, vicenin-2 is, by far, the most abundant flavone derivative, with the highest level in bergamot cultivars (Fantastico, Femminello, and Castaganaro). Additionally, other C-glycosyl flavones (lucenin-2, stellarin-2, isovitexin, scoparin, and lucenin 2,4-methyl ether) are present in significant amounts in the Citrus spp., but in much lower amounts.

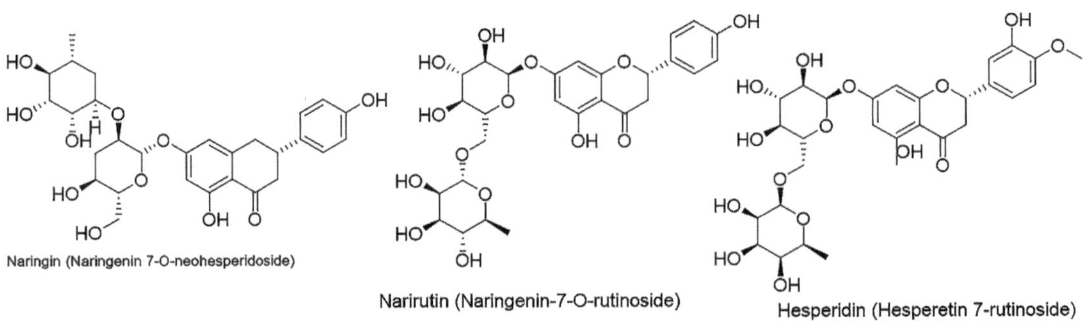

Fig. 1 Chemical structure of flavanones

Rhoifolin, diosmin, and neo diosmin are the main flavone O-glycosides identified with values ranging from ≈2 to ≈30 mg/L. In addition, polymethoxy flavones are present in commonly consumed citrus juices in commercial products but are almost completely absent in the hand-squeezed juice or in the peeled fruit pulp. These lipophilic compounds are mainly located in the peel, where they are components of the essential oil fraction. Flavones (Fig. 2) are also present in the leaves, seeds, and cortex, but the main sources of flavones for human nutrition are the fruits and their juices [3].

Polymethoxyflavones (PMF)

Citrus PMFs are flavones bearing two or more methoxy groups (OCH_3) on their basic benzo-γ-pyrone (15-carbon, C_6–C_3–C_6) skeleton with a carbonyl group at the C_4 position, which have regularity in basic structure. The maximum number of the methoxyl groups can be up to seven, and their A, B, and C rings are sometimes substituted with a hydroxyl group (OH). Nobiletin and tangeretin are the most extensively studied PMFs due to their abundance in citrus plants and their notable pharmacological properties [5] (Fig. 3).

Flavonols

Flavonols are a type of flavonoid characterized by a ketone group. They serve as the foundational components of proanthocyanins and are found in high concentrations in many fruits and vegetables. The most frequently studied flavonols are kaempferol, quercetin, myricetin, and fisetin.

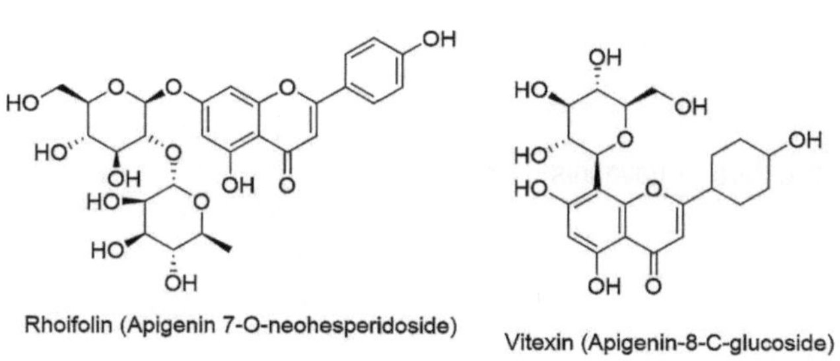

Fig. 2 Chemical structure of flavones

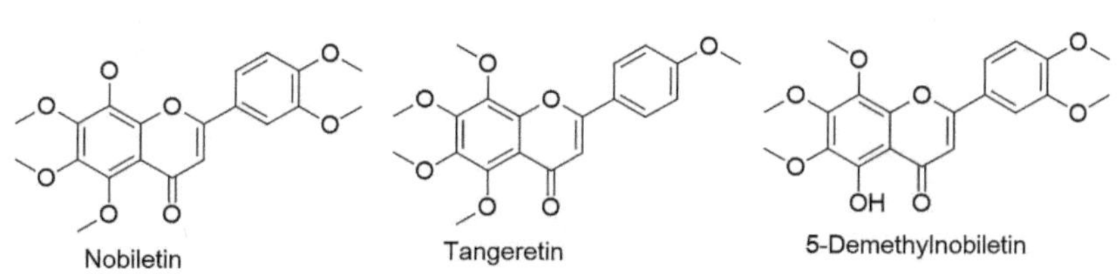

Fig. 3 Chemical structure of polymethoxyflavones (PMF)

In contrast to flavones, flavonols contain a hydroxyl group at position 3 of the C ring (Fig. 4), which may also undergo glycosylation. Similar to flavones, flavonols exhibit considerable diversity in their methylation and hydroxylation patterns. Given the various glycosylation patterns, they are likely the most abundant and largest subgroup of flavonoids found in fruits and vegetables.

Anthocyanins

Anthocyanins are pigments that give color to plants, flowers, and fruits. The most frequently studied anthocyanins include cyanidin, delphinidin, malvidin, pelargonidin, and peonidin. Berries, currants, grapes, and some tropical fruits have high anthocyanin content. The stability and health benefits of anthocyanins make them valuable in various applications within the food industry. The color of anthocyanins is influenced by pH as well as by methylation or acylation at the hydroxyl groups on the A and B rings (Fig. 5).

1.2 Applications in the Food/Non-Food Sector

Citrus-derived flavonoids are essential for regulating plant physiological processes and also offer several health benefits in humans, including cancer prevention, inflammation reduction, protection against atherosclerosis, and defence against neurodegenerative diseases. They have become important resources for the development

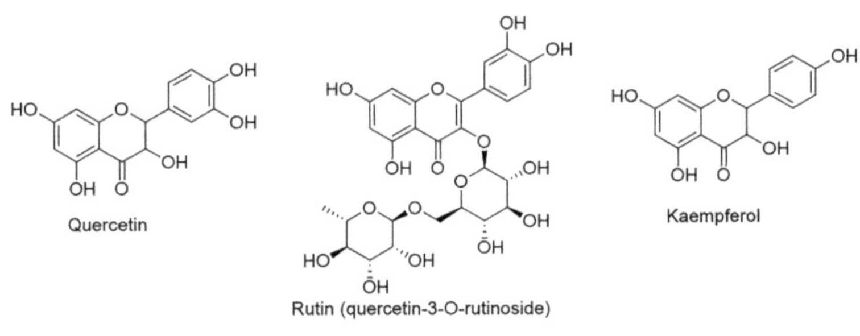

Fig. 4 Chemical structure of flavonols

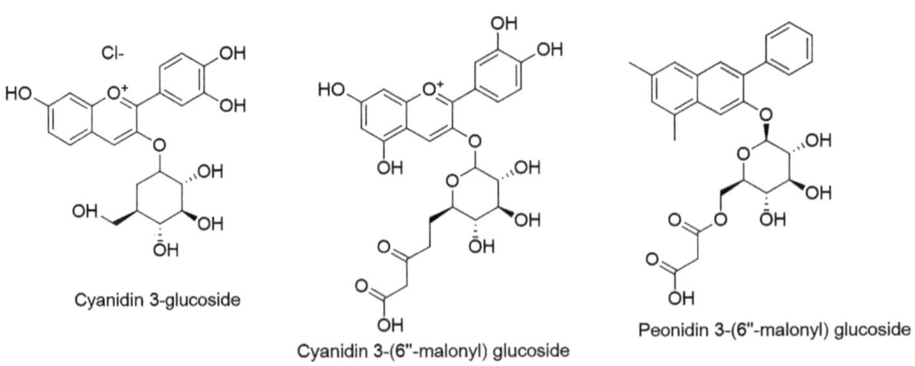

Fig. 5 Chemical structure of anthocyanins

of bioactive compounds in functional food. The health benefits of these food components are closely linked to their chemical structures. The use of citrus by-products in active food packaging provides antioxidant and antimicrobial properties to extend shelf life and maintain quality [4].

2 Materials (Materials/Reagents Required/Steps for Preparation/Isolation/Extraction)

1. Prepare all solutions using ultrapure water and analytical-grade reagents.
2. Prepare and store all reagents at room temperature (unless indicated otherwise).
   - 5% Sodium Nitrite ($NaNO_2$) Solution: Dissolve 5 g of sodium nitrite in 100 mL of water.
   - 10% aluminium nitrite ($Al(NO_3)_3$) (For 100 mL Solution: Dissolve use 10 g of $Al(NO_3)_3$ in 100 mL of water).
   - 1 M Sodium Hydroxide (NaOH): Dissolve 4.0 g of NaOH in water to make 100 mL.
   - 2% Acetic Acid: In about 80–90 mL of distilled water, gradually add the 2 mL of acetic acid to the water while stirring gently. Adjust the volume and bring the total volume to 100 mL.

3 Methods

1. Obtain citrus by-products such as peels, pulp, and seeds from oranges, lemons, grapefruits, limes, mandarin, clementine, etc.
2. Dry the citrus by-products between 40 and 50 °C to constant weight [2, 6, 8].
3. Ground to a powder and store at −30 °C before use.

Solvent Extraction

1. Sonicate (0.25 g/mL) citrus peel in the solvent in a 4:1 (v/v) ethanol–water mixture at 40 °C for 30 min, using a sonication power of 150 W.
2. Remove ethanol from the extracts by evaporation under a vacuum at 40 °C on a rotary evaporator.
3. Then, freeze and lyophilize the sample to remove the water [7].

4 Assessments/Calculations/Formulas

Colorimetric Method (Aluminium Chloride Method)

1. Weigh 2.5 g of a sample, place it in a Soxhlet extractor, and reflux with methanol for more than 12 h at 85 °C.
2. Then evaporate the extract to dryness in a rotary vacuum evaporator at less than 40 °C, and then dissolve in methanol.
3. Add 0.3 mL of 5% $NaNO_2$ to 1 mL extract in a 10 mL volumetric flask and keep the mixture at room temperature for 6 min.
4. Add 0.3 mL of 10% $Al(NO_3)_3$ to the mixture and incubate for 6 min again.
5. Then add 4 mL of 1 M NaOH and methanol to make up to volume.
6. After incubating for 15 min at room temperature for color development, measure absorbance at 510 nm.
7. Total flavonoid content is expressed as rutin equivalents [2, 9].

High-Performance Liquid Chromatography (HPLC)

The flavonoid standards included the following:

1. Phenolic Acids: Caffeic acid, p-coumaric acid, ferulic acid, sinapic acid, and chlorogenic acid.
2. Flavanones: Naringin, hesperidin, and neohesperidin.
3. Flavonols: Quercetin, kaempferol, and rutin.
4. Flavones: Sinensetin, luteolin, and diosmin.
   - Prepare the standards in methanol–dimethyl sulfoxide (DMSO) (v/v; 50:50) and store at −18 °C before use.
   - Mix 0.1 g of total citrus flavonoid sample and 1 mL of methanol–DMSO (v/v; 50:50) in a centrifuge tube, stir for 10 min at room temperature, and centrifuge at 9000 rpm for 15 min at 4 °C.
   - Extract the residues twice with 1 mL of the same extraction solvent.
   - Combine all the supernatants and make up to 5 mL with methanol.
   - Filter the sample solution through a 0.45-μm membrane filter before use.
   - Use an HPLC system with an ultraviolet (UV) detector [2].
   - Carry out the chromatographic separation using a C18 column, 5 μm, 4.0 mm internal diameter (i.d.) × 250 mm.
   - Inject volume 20 μL.

- The mobile phase compose of (A) 2% acetic acid (aqueous) and (B) 0.5% acetic acid (aqueous)-acetonitrile (v/v; 50:50) and gradient elution was performed as follows: 0 min, 95:5; 10 min, 90:10; 40 min, 60:40, 55 min, 45:55; 60 min, 20: 80 and 65 min, 0:100. The mobile phase was filtered under vacuum through a 0.45-μm membrane filter before use.
- Flow rate of 1 mL/min.
- Measure the UV absorbance for phenolic acids, flavanones, and flavonols at 280 nm and flavones at 340 nm [2, 7].
- Maintain the operating temperature at room temperature.

5 Notes

1. NaOH is highly corrosive and can cause severe burns. Always wear gloves and goggles. Add NaOH to water slowly to prevent splashing. Work in a well-ventilated area or under a fume hood.
2. Clearly label the container with the solution concentration and date of preparation.

References

1. Russo M, Bonaccorsi I, Inferrera V, Dugo P, Mondello L (2015) Underestimated sources of flavonoids, limonoids and dietary fiber: availability in orange's by-products. J Funct Foods 12:150–157. https://doi.org/10.1016/j.jff.2014.11.008
2. Wang Y-C, Chuang Y-C, Hsu HW (2008) The flavonoid, carotenoid and pectin content in peels of citrus cultivated in Taiwan. Food Chem 106(1):277–284. https://doi.org/10.1016/j.foodchem.2007.05.086
3. Barreca D et al (2020) Citrus flavones: an update on sources, biological functions, and health promoting properties. Plants 9(3):288. https://doi.org/10.3390/plants9030288
4. Casas Cardoso L, Bastante CC, Serrano CM, de la Ossa EJM (2022) Application of citrus by-products in the production of active food packaging. Antioxidants 11(4):738. https://doi.org/10.3390/antiox11040738
5. Fu Y, Liu W, Soladoye OP (2021) Towards innovative food processing of flavonoid compounds: insights into stability and bioactivity. LWT 150:111968. https://doi.org/10.1016/j.lwt.2021.111968
6. Londoño-Londoño J, de Lima VR, Lara O, Gil A, Pasa TBC, Arango GJ, Pineda JRR (2010) Clean recovery of antioxidant flavonoids from citrus peel: optimizing an aqueous ultrasound-assisted extraction method. Food Chem 119(1):81–87. https://doi.org/10.1016/j.foodchem.2009.05.075
7. Khan MK, Abert-Vian M, Fabiano-Tixier A-S, Dangles O, Chemat F (2010) Ultrasound-assisted extraction of polyphenols (flavanone glycosides) from orange (*Citrus sinensis* L.) peel. Food Chem 119(2):851–858. https://doi.org/10.1016/j.foodchem.2009.08.046
8. Del Río JA, Fuster MD, Gómez P, Porras I, García-Lidón A, Ortuño A (2004) *Citrus limon*: a source of flavonoids of pharmaceutical interest. Food Chem 84(3):457–461. https://doi.org/10.1016/S0308-8146(03)00272-3
9. Li P, Yao X, Zhou Q, Meng X, Zhou T, Gu Q (2022) Citrus peel flavonoid extracts: health-beneficial bioactivities and regulation of intestinal microecology in vitro. Front Nutr 9:888745. https://doi.org/10.3389/fnut.2022.888745
10. National Research Council (US) Committee on Toxicology (1984) Emergency and Continuous Exposure Limits for Selected Airborne Contaminants: Volume 2. Washington (DC): National Academies Press (US). Sodium Hydroxide. Available from: https://www.ncbi.nlm.nih.gov/books/NBK208293/

Chapter 5

Antioxidants in Citrus By-Products

Samandeep Kaur, Parmjit S. Panesar, and Vikrant Singh

Abstract

Citrus by-products, such as peels, seeds, and pulp, are rich in bioactive compounds, including phenolics, flavonoids, and ascorbic acid. These naturally occurring antioxidants play a crucial role in reducing oxidative stress and promoting overall health, making them valuable for applications in the food and nutraceutical industries. To accurately assess their antioxidant potential, various in vitro assays, such as DPPH, FRAP, ABTS, TEAC, and ORAC, are employed, each relying on distinct reaction mechanisms. This chapter explores these analytical methods, outlining their underlying principles, benefits, and limitations to assist researchers in selecting the most suitable approach. A deeper understanding of these techniques is essential for enhancing the utilization of citrus by-products, supporting waste reduction efforts, and advancing sustainable innovations in food and pharmaceutical applications.

Key words Antioxidant activity, Citrus by-products, Reactive oxygen species (ROS), Functional foods, Waste valorization

1 Introduction

Citrus fruits, including oranges (*Citrus sinensis*), lemons (*Citrus limon*), limes (*Citrus aurantifolia*), tangerines (*Citrus reticulata*), and grapefruits (*Citrus paradisi*), are among the most widely cultivated crops worldwide, with their production reaching approximately 166.4 million tons in 2022 [1]. The citrus industry generates large volumes of by-products—such as peels, seeds, pulp residues, pomace, and membranes—which account for more than 50% of the fruit's total weight. Despite being rich in bioactive compounds, these by-products are often discarded, contributing to environmental concerns [2]. However, growing interest in sustainable food processing has highlighted their potential as valuable ingredients in functional foods and nutraceuticals.

Citrus by-products contain a wide range of antioxidant compounds, including flavonoids, carotenoids, ascorbic acid (vitamin C), and phenolic compounds. These bioactives play a crucial role in scavenging free radicals, inhibiting lipid peroxidation, and reducing

Shweta Suri et al. (eds.), *Citrus Fruit Processing By-Products: Technologies and Basic Protocols*, Methods and Protocols in Food Science, https://doi.org/10.1007/978-1-0716-5088-2_5,

Table 1
Common analytical techniques for measuring antioxidant activity of citrus by-products [6]

Technique	Antioxidant capacity assay	Principle of the method	End-product determination
Spectrometry	ORAC (oxygen radical absorbance capacity)	Antioxidant reaction with peroxyl radicals induced by AAPH (*2,2′-azobis-2-amidino-propane*)	Fluorescence decay
Colorimetric assays	CUPRAC (cupric reducing antioxidant capacity)	Reduction of Cu^{2+} to Cu^{+} by antioxidants	Colorimetric change
	FRAP	Reduction of Fe^{3+} complex by antioxidants	Colorimetric change at 593 nm
	ABTS	Antioxidant reaction with ABTS•+ radical cation	Colorimetric change at 734 nm
	DPPH	Antioxidant reaction with DPPH• radical	Colorimetric change at 517 nm

oxidative stress—mechanisms that contribute to the prevention of chronic diseases such as cancer, diabetes, and cardiovascular disorders [3, 4]. Compared to synthetic antioxidants like butylated hydroxytoluene (BHT) and butylated hydroxyanisole (BHA), natural antioxidants derived from citrus waste offer greater stability, efficacy, and safety, making them an attractive alternative for extending shelf life in food products without compromising sensory attributes [5].

To evaluate the antioxidant potential of citrus by-products, various in vitro assays are commonly used, each based on different reaction mechanisms (Table 1). These methods include hydrogen atom transfer (HAT), single electron transfer (SET), and reducing power assays, all of which measure antioxidant activity through distinct pathways. Among the most widely applied techniques are the DPPH (2,2-diphenyl-1-picrylhydrazyl) assay, which assesses radical scavenging ability via electron transfer; the ferric reducing antioxidant power (FRAP) assay, which quantifies the reduction of Fe^{3+} to Fe^{2+}; and the ABTS (2,2′-azino-bis(3-ethylbenzothiazoline-6-sulfonic acid)) assay, which evaluates antioxidant activity in both aqueous and lipid-soluble systems. Additionally, the Trolox equivalent antioxidant capacity (TEAC) assay expresses antioxidant activity relative to Trolox, a vitamin E analog, while the oxygen radical absorbance capacity (ORAC) assay measures fluorescence decay caused by peroxyl radicals, offering a biologically relevant assessment of antioxidant effectiveness [6].

Certain citrus by-products exhibit particularly high antioxidant activity due to their unique composition. Lemon and grapefruit peels are notably rich in polyphenols and flavonoids, which contribute to their strong radical-scavenging properties. Grapefruit peels, in particular, are a significant source of essential oils, with global production reaching approximately 200 tons annually. Lemon peels contain high levels of ascorbic acid and phenolic acids such as ferulic acid, *p*-coumaric acid, and sinapic acid, which enhance their antimicrobial and antioxidant functions. Orange peels, traditionally used in natural remedies, are well known for their immune-boosting properties and their ability to reduce oxidative stress [7].

With their potent antioxidant capacity, citrus by-products present a promising opportunity for food preservation, nutraceuticals, and pharmaceuticals. Their valorization supports sustainable processing strategies and aligns with the circular bioeconomy, reducing food waste while maximizing nutritional and commercial benefits. However, optimizing extraction techniques and refining characterization methods are essential to fully harness their antioxidant potential and ensure their effective incorporation into health-promoting products.

This chapter provides a detailed overview of the extraction methods, quantification techniques, and analytical procedures used to evaluate the antioxidant activity of citrus by-products. A deeper understanding of these methodologies can help researchers develop efficient strategies for utilizing citrus waste in functional food formulations and pharmaceutical applications, ultimately contributing to both human health and environmental sustainability.

2 Materials

Apparatus

Beakers, test-tubes, micropipettes, incubator, and spectrophotometer.

Chemicals and Reagents

DPPH (2,2-diphenyl-1-picrylhydrazyl): 0.1 mM in methanol, methanol, and ethanol (for extraction).

Acetate buffer (300 mM, pH 3.6).

TPTZ (2,4,6-tripyridyl-s-triazine), 10 mM in 40 mM HCl.

Ferric chloride ($FeCl_3.6H_2O$): 20 mM.

ABTS (2,2′-azino-bis-(3-ethylbenzothiazoline-6-sulfonic acid)): 7 mM.

Potassium persulfate: 2.45 mM.

3 Methods

3.1 Sample Preparation

1. Dry citrus peels under controlled conditions (*see* **Note 1**).
2. Grind dried peels using a domestic grinder to obtain a fine powder.
3. Sieve the powder through a 100-mesh screen (*see* **Note 2**).
4. Store the sieved powder in airtight pouches at 4 °C until further use.

3.2 Extract Preparation

1. Weigh 1 g of pulverized citrus by-product powder.
2. Add a known volume of organic solvent (e.g., methanol, ethanol, or a solvent mixture).
3. Stir the mixture occasionally and allow extraction to proceed for 24 h at room temperature (*see* **Notes 3** and **4**).
4. Filter the extract using Whatman No. 42 filter paper or centrifuge to collect the supernatant.
5. Remove excess solvent using a rotary evaporator under reduced pressure and air-dry the remaining residue under a chemical hood (*see* **Notes 5** and **6**).
6. Collect the dried extracts in microfuge tubes and store appropriately for further analysis (*see* **Note 5**).

3.3 Antioxidant Activity Assay Methods

3.3.1 DPPH Radical Scavenging Activity

Principle

The DPPH assay evaluates the electron transfer-based radical scavenging ability of antioxidants. A decrease in absorbance at 517 nm indicates higher antioxidant activity [8].

Procedure

1. Prepare a 0.1 mM DPPH solution in methanol.
2. Mix 1 mL of citrus extract with 3 mL of DPPH solution.
3. Incubate the mixture in the dark at 25 °C for 30 min.
4. Measure absorbance at 517 nm using a spectrophotometer [9].
5. Calculate % scavenging activity using the formula:

$$\text{DPPH Activity}(\%) = \frac{(A_0 - A_1)}{A_0} \times 100$$

where A_0 is the control absorbance and A_1 is the sample absorbance.

Precautions

1. Prepare a fresh DPPH solution as it degrades in light.
2. Ensure uniform mixing for consistent results.
3. Use a blank (solvent only) to correct baseline absorbance.

3.3.2 Ferric Reducing Antioxidant Power (FRAP) Assay

The FRAP assay measures the reduction of Fe^{3+} to Fe^{2+}, indicating the antioxidant power of a sample. The end-product is detected as a colorimetric change at 593 nm.

Principle

Procedure

1. Prepare FRAP reagent by mixing 300 mM acetate buffer (pH 3.6), 10 mM TPTZ (2,4,6-tripyridyl-s-triazine) in 40 mM HCl, and 20 mM $FeCl_3 \cdot 6H_2O$ in a 10:1:1 ratio (*see* **Note** 7).
2. Add 100 μL of citrus extract to 3 mL of FRAP reagent.
3. Incubate at 37 °C for 30 min.
4. Measure absorbance at 593 nm [10].
5. Express antioxidant power as μmol Fe^{2+} equivalents per gram of sample using the equation below:

$$\text{FRAP value} = \frac{(\text{Absorbance of sample} - \text{Absorbance of blank})}{\text{Slope of } Fe^{2+}\text{standard curve}}$$

Precautions

1. Maintain pH 3.6 for optimal reaction conditions.
2. Prepare FRAP reagent fresh before each use.
3. Avoid contamination with reducing agents.

3.3.3 ABTS Radical Scavenging Assay

The ABTS assay evaluates antioxidant capacity by measuring the color change of $ABTS^{\bullet +}$ radicals in the presence of antioxidants at 734 nm.

Principle

Procedure

1. Prepare ABTS stock solution (7 mM).
2. Generate ABTS•+ radical by mixing ABTS with 2.45 mM potassium persulfate and incubating in the dark for 12–16 h.
3. Dilute the ABTS•+ solution with ethanol to an absorbance of 0.70 ± 0.02 at 734 nm.
4. Mix 1 mL of citrus extract with 2 mL of diluted ABTS•+ solution.
5. Incubate for 6 min at room temperature.
6. Measure absorbance at 734 nm and express results as Trolox equivalents (TEAC, μmol Trolox/g sample) [11].
7. Calculate % scavenging activity using the formula:

$$\text{ABTS Scavenging Activity}(\%) = \frac{(A_0 - A_1)}{A_0} \times 100$$

where A_0 is the control absorbance and A_1 is the sample absorbance.

Precautions

1. Use freshly prepared ABTS•+ solution.
2. Standardize ABTS absorbance to 0.70 ± 0.02 before testing.
3. Keep samples in the dark to prevent radical degradation.

3.3.4 Oxygen Radical Absorbance Capacity (ORAC) Assay

Principle

The ORAC assay evaluates antioxidant capacity by measuring the ability of a sample to scavenge peroxyl radicals (ROO•) generated by 2,2′-azobis(2-amidinopropane) dihydrochloride (AAPH). The reaction is monitored through the fluorescence decay of fluorescein, where the degree of protection against oxidative degradation reflects antioxidant potential.

Procedure

1. Prepare fluorescein solution (10 nM) in 75 mM phosphate buffer (pH 7.4).
2. Prepare AAPH solution (150 mM) in phosphate buffer (*see* **Note 7**).
3. Prepare Trolox standard solutions (10–100 μM) in phosphate buffer.
4. Add 150 μL of fluorescein solution to a 96-well microplate.
5. Add 25 μL of the citrus extract or Trolox standard.
6. Incubate at 37 °C for 10 min in the dark.
7. Add 25 μL of freshly prepared AAPH solution to initiate oxidation.
8. Immediately measure fluorescence every minute for 90 min at excitation/emission wavelengths of 485/535 nm using a fluorescence spectrophotometer [12].
9. Calculate the area under the curve (AUC) for each sample using the formula:

$$\text{Net AUC} = (\text{AUC Sample} - \text{AUC blank})$$

10. Express antioxidant capacity as Trolox equivalents (μmol Trolox/g sample) using the Trolox standard curve.

Precautions

1. Use freshly prepared AAPH solution, as it degrades rapidly.
2. Protect the fluorescein solution from light to prevent photodegradation.
3. Run all samples in triplicate for accurate results.

3.3.5 Cupric Reducing Antioxidant Capacity (CUPRAC) Assay

Principle

The CUPRAC assay determines antioxidant reducing power by measuring the reduction of Cu^{2+} to Cu^{+} in the presence of antioxidants. The resulting Cu^{+}–neocuproine complex produces an intense orange color, detectable at 450 nm.

Procedure

1. Prepare CUPRAC reagent by mixing: 10 mM $CuCl_2$ (copper (II) chloride), 7.5 mM neocuproine (Nc) in ethanol, and 1 M ammonium acetate buffer (pH 7.0) in a 1:1:1 ratio (*see* **Note 7**).
2. Prepare Trolox standard solutions (10–100 μM) in ethanol.
3. Add 1 mL of citrus extract or Trolox standard to 1 mL of CUPRAC reagent.
4. Incubate the mixture at 25 °C for 30 min.
5. Measure absorbance at 450 nm using a spectrophotometer [13].
6. Calculate antioxidant capacity as Trolox equivalents (μmol Trolox/g sample) using the Trolox standard curve, using the formula:

$$\text{CUPRAC value} = \frac{(\text{Absorbance of sample} - \text{Absorbance of blank})}{\text{Slope of Trolox standard curve}}$$

Precautions

1. Prepare fresh CUPRAC reagent before each assay.
2. Ensure proper pH control (pH 7.0) for reproducible results.
3. Conduct blank measurements with solvent alone to correct for background absorbance.

4 Discussion and Conclusion

The assessment of antioxidant activity in citrus by-products is critical for understanding their potential applications in food preservation, nutraceuticals, and pharmaceuticals. This study evaluated the most commonly used antioxidant assays—DPPH, FRAP, ABTS, TEAC, ORAC, and CUPRAC—each offering distinct advantages and limitations depending on the reaction mechanism involved. While DPPH and ABTS assays effectively measure free radical scavenging capacity, FRAP and CUPRAC provide insights into reducing power. ORAC, in contrast, assesses peroxyl radical scavenging based on fluorescence decay, making it more relevant for biological systems. The TEAC assay standardizes antioxidant potential against Trolox, facilitating comparative studies across different matrices.

Given the complexity of citrus by-products, which contain both hydrophilic and lipophilic antioxidants, a single assay may not fully capture their overall antioxidant capacity. Instead, a combination of assays is recommended to obtain a more comprehensive evaluation. The choice of method should also consider factors such as solvent compatibility, sample matrix, and assay sensitivity. These considerations are essential for optimizing extraction techniques and ensuring accurate quantification of bioactive compounds.

The valorization of citrus by-products is not only a means of reducing food industry waste but also aligns with sustainable processing strategies and the principles of a circular bioeconomy. By refining antioxidant extraction and assessment methodologies, these by-products can be repurposed into high-value ingredients with functional health benefits. Future research should focus on improving extraction efficiency, standardizing antioxidant assessment protocols, and investigating the stability and bioavailability of citrus-derived antioxidants in various food and pharmaceutical applications. Expanding knowledge in these areas will further enhance the utilization of citrus by-products, transforming them from agricultural waste into valuable resources for health-promoting applications.

5 Notes

1. Drying can be performed using sun drying, oven drying, or tray drying, etc. The objective is to reduce water activity while preserving nutritional content.
2. You can also modify the particle size to evaluate its impact on the extraction efficiency of bioactive compounds from the matrix.
3. For citrus antioxidant extraction, the choice of solvent depends on the antioxidant's polarity—polar solvents (e.g., water, ethanol, and methanol) are suitable for hydrophilic antioxidants like vitamin C and flavonoids, while nonpolar solvents (e.g., hexane and ethyl acetate) are ideal for lipophilic antioxidants such as carotenoids and limonoids; a mixture of solvents can be used to extract a broader range of compounds.
4. Further, the extraction conditions depend on the chosen extraction system. For example, in maceration and Soxhlet extraction, variables such as temperature, extraction time, solvent concentration, and sample-to-solvent ratio can be adjusted. Similarly, these parameters can also be modified in ultrasound-assisted and microwave-assisted extraction. Additionally, for ultrasound-assisted extraction, factors like power

and pulse can be varied, while for microwave-assisted extraction, microwave power can also be adjusted.

5. Maintain proper labeling and storage conditions for all samples.
6. Ensure complete solvent evaporation to avoid contamination.
7. Use appropriate safety precautions while handling organic solvents.

References

1. FAOSTAT (2023) Crops and livestock products. Food and Agriculture Organization of the United Nations (FAO). Accessed from: https://www.fao.org/faostat/en/#data/QCL
2. Kaur S, Panesar PS, Chopra HK (2023) Citrus processing by-products: an overlooked repository of bioactive compounds. Crit Rev Food Sci Nutr 63(1):67–86
3. Andrade MA, Barbosa CH, Shah MA, Ahmad N, Vilarinho F, Khwaldia K et al (2022) Citrus by-products: valuable source of bioactive compounds for food applications. Antioxidants 12(1):38
4. Suri S, Singh A, Nema PK (2023) Bioactive compounds in citrus fruits: extraction and identification. In: Recent advances in citrus fruits. Springer International Publishing, Cham, pp 427–450
5. Petcu CD, Tăpăloagă D, Mihai OD, Gheorghe-Irimia RA, Negoiță C, Georgescu IM et al (2023) Harnessing natural antioxidants for enhancing food shelf life: exploring sources and applications in the food industry. Foods 12(17):3176
6. Munteanu IG, Apetrei C (2021) Analytical methods used in determining antioxidant activity: a review. Int J Mol Sci 22(7):3380
7. Saleem M, Durani AI, Asari A, Ahmed M, Ahmad M, Yousaf N, Muddassar M (2023) Investigation of antioxidant and antibacterial effects of citrus fruits peels extracts using different extracting agents: phytochemical analysis with in silico studies. Heliyon 9(4):e15433
8. Brand-Williams W, Cuvelier ME, Berset CLWT (1995) Use of a free radical method to evaluate antioxidant activity. LWT-Food Sci Technol 28(1):25–30
9. Su MS, Shyu YT, Chien PJ (2008) Antioxidant activities of citrus herbal product extracts. Food Chem 111(4):892–896
10. Benzie IF, Strain JJ (1996) The ferric reducing ability of plasma (FRAP) as a measure of "antioxidant power": the FRAP assay. Anal Biochem 239(1):70–76
11. Re R, Pellegrini N, Proteggente A, Pannala A, Yang M, Rice-Evans C (1999) Antioxidant activity applying an improved ABTS radical cation decolorization assay. Free Radic Biol Med 26(9–10):1231–1237
12. Ou B, Hampsch-Woodill M, Prior RL (2001) Development and validation of an improved oxygen radical absorbance capacity assay using fluorescein as the fluorescent probe. J Agric Food Chem 49(10):4619–4626
13. Apak R, Güçlü K, Özyürek M, Çelik SE (2008) Mechanism of antioxidant capacity assays and the CUPRAC (cupric ion reducing antioxidant capacity) assay. Microchim Acta 160:413–419

Chapter 6

Dietary Fibers in Citrus By-Products

Shweta Suri and Anupama Singh

Abstract

Citrus by-products, such as peels, seeds, and pomace, serve as significant sources of dietary fiber, encompassing both soluble and insoluble fibers. Citrus peels, particularly those derived from oranges, lemons, and other citrus fruits, are notably high in dietary fiber content. This chapter delves into the enzymatic–gravimetric method, which provides a reliable approach for quantifying the total dietary fiber present in citrus peels. These peels represent a valuable source of both soluble and insoluble fibers, which may confer various health benefits. The enzymatic process employed includes the use of heat-stable α-amylase, followed by digestion with protease and amyloglucosidase to effectively remove protein and starch. Subsequently, the aqueous digest is diluted with four volumes of ethanol to precipitate the soluble dietary fiber. A comprehensive understanding of the components of dietary fiber found in citrus, as well as the estimation process, is essential for enhancing the application of citrus by-products in various fields.

Key words Dietary fiber, Citrus by-products, Soluble fiber, Roughage, Waste valorization

1 Introduction

Citrus fruits represent one of the largest families of fruits cultivated globally. China, Brazil, and India are the leading producers of citrus fruits, collectively accounting for 49.22% of total global production [1]. These fruits are widely appreciated for their appealing taste, aromatic qualities, and notable nutritional benefits. Rich in nutrients, phytochemicals, and bioactive polyphenols, citrus fruits possess significant antioxidant potential, which contributes to their protective effects on human health [2, 3].

In addition to the edible portions of citrus fruits, the nonedible components, particularly the peels (which constitute approximately 50–55% of the total fruit weight), represent a significant source of valuable biologically active compounds. These compounds include phenols, flavonoids, essential oils, carotenoids, organic acids, ascorbic acid, and various vitamins [4]. The waste generated from citrus fruits—comprising peels (flavedo and albedo), seeds, and pomace—is increasingly being recognized as a valuable resource

Shweta Suri et al. (eds.), *Citrus Fruit Processing By-Products: Technologies and Basic Protocols*, Methods and Protocols in Food Science, https://doi.org/10.1007/978-1-0716-5088-2_6,

for various applications. Notably, citrus peels serve as an excellent reservoir of antioxidant compounds, pectin, and dietary fiber, all of which contribute to the reduction of blood glucose and cholesterol levels [5, 6].

According to the American Association of Cereal Chemists (AACC), dietary fiber consists of the remnants of the edible portions of plants, along with analogous carbohydrates that resist digestion and absorption in the human small intestine, undergoing complete or partial fermentation in the large intestine. These dietary fibers are composed of polysaccharides, oligosaccharides, cellulose, lignin, and related plant substances [7]. Additionally, the United States Food and Drug Administration (FDA) defines dietary fiber as nondigestible soluble and insoluble carbohydrates (with three or more monomeric units) and lignin that are intrinsic and intact within plants, as well as isolated or synthetic nondigestible carbohydrates (of three or more monomeric units) that the FDA has determined to confer physiological benefits to human health [8]. Dietary fiber represents a complex mixture of chemically diverse compounds, including cellulose, hemicelluloses, pectin, lignin, and gums. It is categorized into two types: soluble fibers, which dissolve in water to form a gel, and insoluble fibers, which do not dissolve in water and pass through the digestive system largely unchanged [9].

Fiber extracted from citrus by-products serves as a significant source of both soluble and insoluble fiber, recognized for its potential nutritional and health benefits [10]. Notably, the ratio of soluble to insoluble dietary fiber in citrus peels is superior to that found in cereals. For instance, lemon peels have been reported to contain a total dietary fiber content of 89.67% [11]. Furthermore, studies indicate that the dietary fiber content in the peels of oranges, grapefruits, and limes ranges from 44.2% to 68.3%, with orange peels specifically measured at 64.3% [12].

In addition to their nutritional value, dietary fibers are known to enhance the physical, chemical, and functional properties of various food products. In baked goods, for example, dietary fibers can modify texture, increase volume, extend shelf life, and serve as a viable substitute for wheat flour, thereby improving overall nutritional quality [13]. Recent research has focused on the use of dietary fiber derived from nonconventional sources, such as horticultural by-products like citrus peels and pomace, in the development of food products. This approach not only aims to utilize industrial by-products effectively but also promotes the creation of economic and sustainable products [14].

To assess the dietary fiber content in citrus by-products, enzymatic gravimetric assays are frequently employed. This methodology involves subjecting the sample to enzymatic treatments that replicate the digestive processes occurring in the human small intestine, thereby enabling the quantification of both soluble and insoluble fiber fractions.

Typically, the enzymatic gravimetric method commences with the treatment of citrus peels using heat-stable α-amylase, facilitating the breakdown of starch. This is followed by the application of protease to digest proteins, while amyloglucosidase is introduced to hydrolyze any remaining sugars. The combination of heat-stable α-amylase, protease, and amyloglucosidase effectively eliminates protein and starch content. Subsequently, the aqueous digest is diluted with four volumes of ethanol to precipitate soluble dietary fiber. The resultant residue is then filtered and washed successively with 78% ethanol, 95% ethanol, and acetone, before being dried and weighed. For analytical purposes, one duplicate of the sample is analyzed for protein content, while the other is incinerated at 525 ° C to ascertain the ash content. The total dietary fiber obtained is subsequently adjusted for method blanks, which account for the parameters of protein and ash determinations [15]. Citrus by-products, characterized by their high dietary fiber content, represent a significant opportunity for application in the development of food products, nutraceuticals, and pharmaceuticals. The valorization of these by-products not only supports sustainable processing strategies but also aligns with the principles of the circular bioeconomy, effectively reducing food waste while enhancing both nutritional and commercial advantages.

This chapter offers a comprehensive overview of the enzymatic gravimetric method, along with quantification techniques and analytical procedures employed to assess the dietary fiber content of citrus by-products. An in-depth understanding of these methodologies can empower researchers to devise effective strategies for the utilization of citrus waste in functional food formulations and pharmaceutical applications, thereby contributing to improved human health and environmental sustainability.

2 Materials

Apparatus

Beakers, test-tubes, micropipettes, incubator, Kjeldahl apparatus, muffle furnace, vacuum filter, porcelain crucible, hot air oven, burner, and digital balance.

Chemicals and Reagents

Sodium Phosphate Dibasic: 0.1 mM in methanol.

Sodium Phosphate Monobasic Monohydrate: 300 mM (pH 3.6).

Methanol.

Ethanol.

α-Amylase.

Amyloglucosidase.

Protease.

Hydrochloric acid.

Sulfuric acid.

Copper (II) sulfate pentahydrate.

Potassium sulfate.

Boric acid.

Mixed indicators-bromocresol green and methyl red.

Acetone.

Whatman No. 54 filter paper.

3 Methods

3.1 Sample Preparation

1. Clean the citrus peels (the outer layer and the white part) by washing them with safe drinking water to remove impurities, dust, dirt, and pesticide residues.
2. Use a clean cloth to dry off any water left on the peels before you start drying them.
3. Dry the citrus peels in a hot air dryer set to 60 °C with an air speed of 2 m/s for 4–5 h.
4. Use a laboratory processor to grind the dried peels into a fine powder.
5. Sift the powder through a 60-mesh screen (250 μm).
6. Store the citrus peel powder in airtight zip-lock bags at 4 °C for dietary fiber analysis.

3.2 Sample Extract Preparation (Aqueous Extraction)

1. Weigh 10 g of citrus peel powder accurately.
2. Add 40 mL of deionized water.
3. Stir the mixture with a magnetic stirrer and let it sit at room temperature to extract dietary fiber.
4. Heat the mixture to 60 °C and keep it there for 60 min.
5. Pour the extracts into Falcon tubes and store them properly for future analysis.

3.3 Dietary Fiber Estimation Method

Principle

The estimation of dietary fiber involves using heat-stable α-amylase, which digests the sample in conjunction with protease and amyloglucosidase to eliminate protein and starch. The resulting aqueous digest is then diluted with four volumes of ethanol to precipitate the soluble dietary fiber. After precipitation, the residue is filtered and washed with 78% ethanol, 95% ethanol, and acetone. It is subsequently dried and weighed. One duplicate sample is analyzed for protein content, while the other is incinerated at 525 °C to determine the ash content [7, 16].

Procedure

1. Add 1 mL of citrus peel extract to 50 μL of heat-stable α-amylase.
2. Heat the mixture to 100 °C for 15 min.
3. Let the mixture cool and adjust the pH to 7.5 using HCl.
4. Add 100 μL of protease to the mixture and heat it to 60 °C for 30 min.
5. Cool the mixture again and adjust the pH to 4.5 using HCl.
6. Add 200 μL of amyloglucosidase to the mixture and heat it to 60 °C for another 30 min.
7. Allow the mixture to cool and then centrifuge it carefully.
8. Collect the wet pellet and the liquid (supernatant) separately.
9. Follow the next steps to separate the soluble and insoluble dietary fiber by filtering.
10. Calculate the percentage of total dietary fiber using the equation below:

$$\text{Total Dietary Fiber}(\%) = \frac{\text{Weight of SDF} + \text{Weight of IDF} - \text{Protein content} - \text{Ash content}}{\text{Intital weight of citrus peel powder}} \times 100$$

Precautions

1. Use high-quality enzymes like amylase, protease, and amyloglucosidase that specifically target the components of citrus by-products.
2. Prepare a fresh solution for the enzymes.
3. Mix and stir the reagents and enzymes evenly for consistent results.

3.3.1 Insoluble Dietary Fiber

Principle

Insoluble dietary fiber is a type of fiber that does not dissolve in water and plays a crucial role in promoting digestive health. It assists in moving food through the digestive system, adds bulk to stools, and may help in the prevention and management of constipation. This type of fiber is found in the outer layers of various plant-based foods, including the skins of citrus fruits and vegetables, as well as in whole grains, nuts, and seeds.

Procedure

1. Rinse the wet pellet obtained after centrifugation with distilled water.
2. Use Whatman filter paper to filter the residue.
3. Dry it in a hot-air oven at 60 °C.
4. Collect the insoluble fiber and weigh it on a calibrated balance.

3.3.2 Soluble Dietary Fiber

Principle

Soluble fiber is a type of dietary fiber that dissolves in water, forming a gel-like substance in the digestive tract. It can help lower cholesterol levels, stabilize blood sugar, and improve digestion. The method for estimating soluble dietary fiber involves enzymatic digestion followed by gravimetric measurement. After treating the food sample with enzymes, the soluble fiber fraction is separated and quantified.

Procedure

1. Heat 95% ethanol to 60 °C for 10 min.
2. Wash the supernatant with the preheated 95% ethanol to help the soluble dietary fiber settle.
3. Collect the solid material and wash it again with 20 mL of 78% ethanol, 10 mL of 95% ethanol, and 10 mL of acetone.
4. Filter the solid using Whatman filter paper and dry it in a hot-air oven at 60 °C.
5. Gather the soluble fiber and weigh it using a calibrated balance.

3.3.3 Protein Content

Principle

The protein content of soluble and insoluble dietary fiber residues is determined using the Kjeldahl method. In this process, the sample is oxidized in the presence of sulfuric acid, which facilitates the conversion of nitrogenous compounds into ammonium sulfate. Ammonia is subsequently liberated by adding an excess of alkali and is quantitatively distilled into a measured volume of standard hydrochloric or sulfuric acid. The acid, which is not neutralized, is then back-titrated with a standard alkali to ascertain the nitrogen content of the sample. To calculate the protein content, the total nitrogen is typically multiplied by an appropriate conversion factor, which is based on the percentage of nitrogen present in the specific protein being analyzed.

Procedure

To digest samples for protein estimation, follow these steps:

1. Weigh a known amount of dietary fiber residue. Add 10 g of a digestion mixture made of 96% copper (II) sulfate pentahydrate and 4% potassium sulfate, along with 25 mL of concentrated sulfuric acid (18 mol/L) in a Kjeldahl flask.
2. Heat the flask for 2–3 h until the mixture turns a light green-blue color, indicating it is carbon-free.
3. Allow the digested sample to cool. If it starts to solidify, add a small amount of water and gently swirl to mix.
4. Take 20 mL of the digested sample and distill it with 40% sodium hydroxide solution for 15–20 min.
5. Collect the ammonia that is released in a separate conical flask. Add 25 mL of 4% boric acid and a few drops of mixed indicator (bromocresol green and methyl red in a 2 to 1 ratio).
6. Fill a burette with 0.1 N sulfuric acid and titrate the distillate until the solution turns light pink.
7. Record the reading when the light pink color is reached.
8. Perform a blank test using sucrose instead of the test sample.

Calculate the protein (%) using the following equations:

$$\text{Nitrogen }(\%) = \frac{\text{Sample titre volume} - \text{Blank titre volume}}{\text{Sample weight}}$$

$$\times 0.0014 \times \text{total vol.of sample aliquot distilled} \times 100$$

$$\text{Protein }(\%) = \text{Nitrogen }(\%) \times \text{conversion factor }(6.25)$$

3.3.4 Ash Content

Principle

Ash content refers to the inorganic residue or mineral content that remains after the ignition or complete oxidation of organic matter in a food sample. To determine the ash content, the sample is incinerated in a muffle furnace at a temperature of 500 ± 50 °C, which is sufficient to burn off all organic materials. The ash content is expressed as a percentage based on the difference in weight before and after the ashing process.

Procedure

1. Weigh a known amount of dry dietary fiber residue (to the nearest 0.0001 g) and put it into a pre-weighed and dried porcelain crucible.
2. Place the crucibles into a muffle furnace.

3. Heat the samples at 550 °C until no carbon is left and only white or gray ash remains.
4. Turn off the furnace and wait 2 h before opening it so it can cool down.
5. Move the crucibles directly to a desiccator and let them cool to room temperature.
6. Weigh the crucibles again and record the results.

Calculate the ash content using the following equation:

$$\text{Ash content } (\%) = \frac{\text{Weight after ashing} - \text{Weight of empty crucible}}{\text{Sample weight}} \times 100$$

3.4 Discussion and Conclusion

Functional compounds, such as dietary fiber obtained from citrus by-products, hold significant promise for the food industry due to their multifaceted benefits. These compounds not only contribute bulk to diets, which can aid in satiety and overall digestive health, but they also play a crucial role in slowing down oxidative changes in food products, thereby extending shelf life and maintaining quality. Furthermore, they enhance the nutritional profile of food, providing essential nutrients and promoting overall wellness.

Evaluating the dietary fiber content in citrus by-products, such as orange peels, lemon rinds, and lime waste, is essential for identifying their potential applications in food formulation and processing. This chapter presents a comprehensive overview of the methodologies employed to quantify insoluble, soluble, and total dietary fiber content in these by-products.

One of the most reliable techniques for assessing dietary fiber content is the enzymatic method. This approach simulates the human digestive process through a carefully devised digestion procedure. It involves the use of specific enzymes, including heat-stable α-amylase for starch breakdown, protease for protein digestion, and amyloglucosidase to completely hydrolyze residual starch into glucose. By replicating these biological processes in a controlled laboratory setting, researchers can more accurately determine the dietary fiber content in citrus by-products, paving the way for their effective utilization in the food sector. This enhanced understanding may lead to innovative food product developments that capitalize on the nutritional and functional attributes of dietary fiber derived from citrus sources.

Increasing knowledge in the extraction and assessment of dietary fiber will further boost the use of citrus by-products, turning them from agricultural waste into valuable resources for health-enhancing applications.

4 Notes

1. Take duplicate or triplicate measurements and calculate the average to reduce errors.
2. When analyzing dietary fiber from citrus by-products, it is important to address any chemicals that might interfere with the results.
3. The ash and protein content can affect the accuracy of dietary fiber tests, so make sure to measure and consider them.
4. Use proper purification methods to remove any lipids from citrus by-products.
5. Citrus by-products have both soluble fiber (like pectin) and insoluble fiber (like cellulose and hemicellulose). You need to separate and measure these types accurately.
6. Follow safety protocols when using concentrated acids and other reagents.

References

1. FAOSTAT (2023) Crops and livestock products. Food and Agriculture Organization of the United Nations (FAO). Accessed from: https://www.fao.org/faostat/en/#data/QCL
2. Suri S, Singh A, Nema PK (2021) Recent advances in valorization of citrus fruits processing waste: a way forward towards environmental sustainability. Food Sci Biotechnol 30(13): 1601–1626
3. Kaur S, Panesar PS, Chopra HK (2023) Citrus processing by-products: an overlooked repository of bioactive compounds. Crit Rev Food Sci Nutr 63(1):67–86
4. Suri S, Singh A, Nema PK, Taneja NK (2022) A comparative study on the debittering of Kinnow (Citrus reticulate L.) peels: microbial, chemical, and ultrasound-assisted microbial treatment. Fermentation 8(8):389
5. Kumari M, Singh S, Chauhan AK (2023) A comparative study of the extraction of pectin from Kinnow (Citrus reticulata) peel using different techniques. Food Bioprocess Technol 16(10):2272–2286
6. Suri S, Singh A, Nema PK, Malakar S, Arora VK (2022) Sweet lime (Citrus limetta) peel waste drying approaches and effect on quality attributes, phytochemical and functional properties. Food Biosci 48:101789
7. AACC International. Approved methods of analysis, 11th edn. Method 32-05.01. Total dietary fiber. Approved October 16, 1991. Method 32-07.01. Soluble, insoluble, and total dietary fiber in foods and food products. Approved October 16, 1991. Method 32-20.01. (PDF) Measurement of Total Dietary Fiber Using AOAC Method 2009.01 (AACC International Approved Method 32-45.01): Evaluation and Updates
8. Turner ND, Lupton JR (2021) Dietary fiber. Adv Nutr 12(6):2553–2555
9. Siddiqui H, Sultan Z, Yousuf O, Malik M, Younis K (2023) A review of the health benefits, functional properties, and ultrasound-assisted dietary fiber extraction. Bioact Carbohydr Diet Fibre 30:100356
10. Berk Z (2016) By-products of the citrus processing industry. In: Berk Z (ed) Citrus fruit processing. Academic Press, San Diego, pp 219–233
11. Zhang Y, Liao J, Qi J (2020) Functional and structural properties of dietary fiber from citrus peel affected by the alkali combined with high-speed homogenization treatment. Lwt 128: 109397
12. Figuerola F, Hurtado ML, Estévez AM, Chiffelle I, Asenjo F (2005) Fibre concentrates from apple pomace and citrus peel as potential fibre sources for food enrichment. Food Chem 91(3):395–401
13. Guldiken B, Gulsunoglu Z, Bakir S, Catalkaya G, Capanoglu E, Nickerson M (2021) Innovations in functional foods development. In: Food technology disruptions. Academic Press, London, pp 73–130

14. Mahmood MW, Abraham-Nordling M, Håkansson N, Wolk A, Hjern F (2019) High intake of dietary fibre from fruit and vegetables reduces the risk of hospitalisation for diverticular disease. Eur J Nutr 58(6):2393–2400
15. Pereira A, Añibarro-Ortega M, Nogueira A, Barros L, Pinela J (2023) Optimization of dietary fiber extraction from quince peel. Eng Proc 37(1):78. https://doi.org/10.3390/ECP2023-14690
16. Kaur S, Panesar PS, Chopra HK (2023) Extraction of dietary fiber from kinnow (Citrus reticulata) peels using sequential ultrasonic and enzymatic treatments and its application in development of cookies. Food Biosci 54: 102891

Chapter 7

Quantification of Carotenoids in Citrus By-Products

Divya Arora, Sanjana Kumari, and Barjinder Pal Kaur

Abstract

Carotenoids of citrus by-products are utilized in both food and nonfood industries. Their quantification in citrus by-products depends on sample type (simple or complex), extraction process (solvent, saponification, ultrasound, microwave, etc.), and suitable quantification method (HPLC, MS, spectrophotometry, etc.). Combining the extraction and quantification methods determines the precision of carotenoid separation, identification, and quantification. This chapter focuses on different techniques and procedures for quantifying major carotenoids: β-cryptoxanthin, ζ-carotene, zeaxanthin, ζ-carotene, and violaxanthin. The primary methods for quantification include HPLC, MS, spectrophotometer, NIR, FTIR, and GC. These quantified carotenoids offer valuable antioxidants and nutritional potential for functional foods and packaging applications.

Key words Citrus waste, Citrus by-products, Carotenoids, HPLC, Spectrophotometry

1 Introduction

1.1 Carotenoids Found in Citrus Fruit By-Products

Citrus fruits, particularly oranges and limes, produce considerable waste from the juice industry. After processing, approximately 50–70% of the citrus fruit mass becomes a by-product, contributing to the global production of approximately 15 million tons of by-products per annum. Valorizations of these by-products are crucial as they are primarily dumped into landfills, leading to the generation of methane gas, which contributes to global warming 25 times more than CO2 [1, 2]. Researchers are now exploring different food and nonfood applications of citrus fruit by-products, as shown in Fig. 1. The citrus by-products, including peel, pulp, and seeds, possess a substantial concentration of carotenoids—naturally occurring pigments recognized for providing vivid orange, yellow, and red colors to fruit tissues and offering considerable health advantages as antioxidants.

The amount and type of carotenoids present in the citrus fruits depend on several factors, including the types of citrus fruits, their variety and cultivars, geographical and climatic conditions,

Shweta Suri et al. (eds.), *Citrus Fruit Processing By-Products: Technologies and Basic Protocols*, Methods and Protocols in Food Science, https://doi.org/10.1007/978-1-0716-5088-2_7,

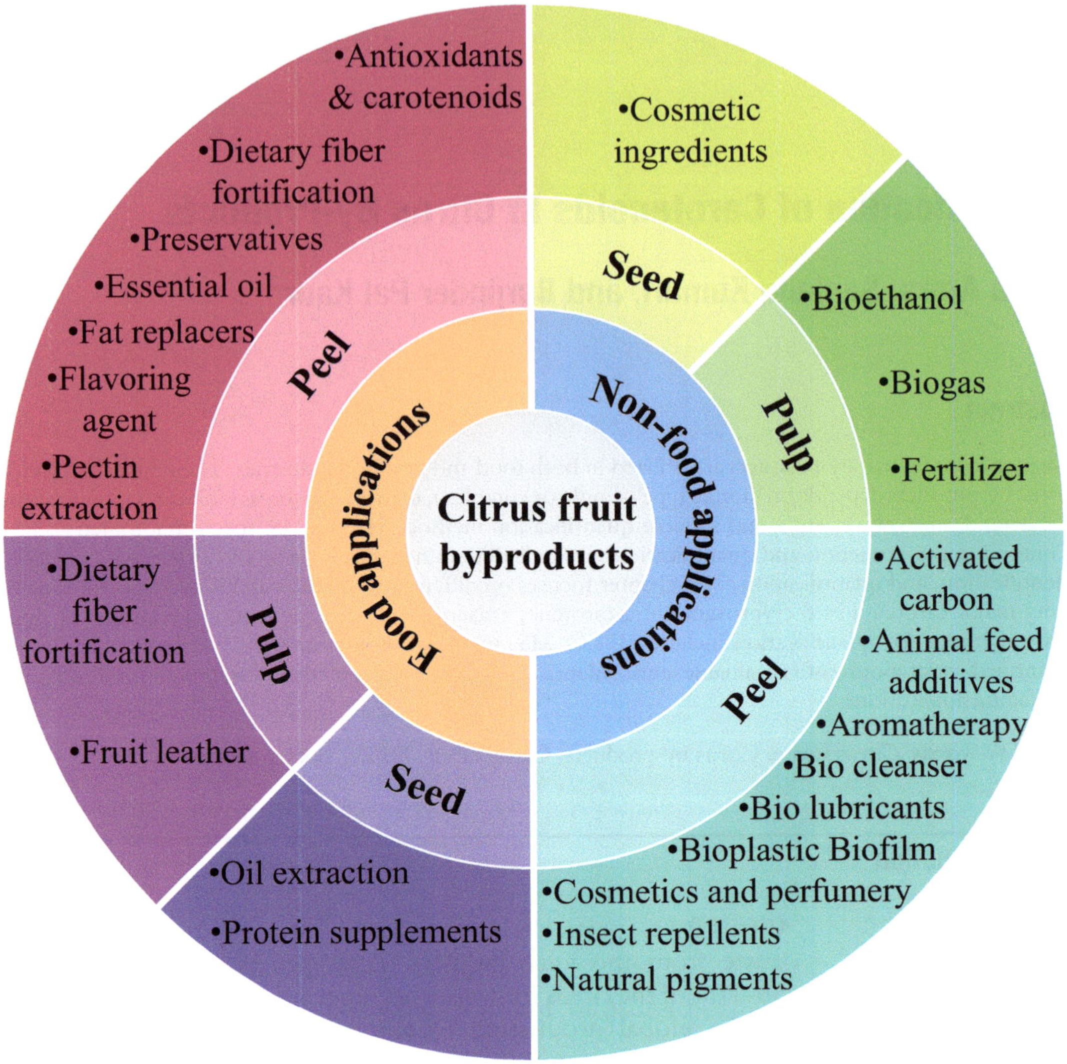

Fig. 1 Wheel chart showing applications of different citrus fruit by-products

maturation or ripening stage, handling during harvesting, and postharvest technologies. Carotenes, or hydrocarbon carotenoids, and xanthophylls, or oxygenated derivatives, comprise the two main categories of the over 600 recognized carotenoids [3]. Some of the most common carotenoids found in the citrus by-products are listed in Table 1.

1.2 Applications in the Food/Nonfood Sector

Carotenoids extracted from citrus by-products such as peels, pulp, or seeds contain valuable bioactive compounds that can be used as antioxidants, colorants, supplements, or ingredients. The green mass in plant material can be utilized for carotenoid extraction and may serve as a food additive, a biological supplement in cosmetics, medicine, and veterinary applications [4]. Apricot kernel

Table 1
Different citrus fruit by-products and the carotenoid groups present, along with their chemical structure and names

Carotenoid group	Carotenoid name (trivial name)	Semi-systematic name	Citrus fruit source	Chemical structure
Carotenes	α-Carotene	β,ε-Carotene	Orange; grapefruit	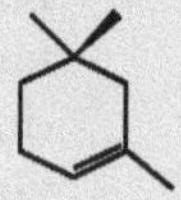
	β-Carotene	β,β-Carotene	Orange; grapefruit; pumello; lemon	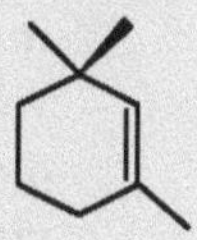
	δ-Carotene	ε,ψ-Carotene		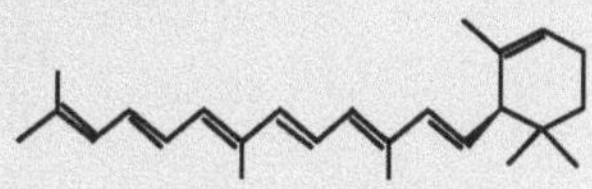
	Lycopene	ψ,ψ-Carotene	Grapefruit; pumello	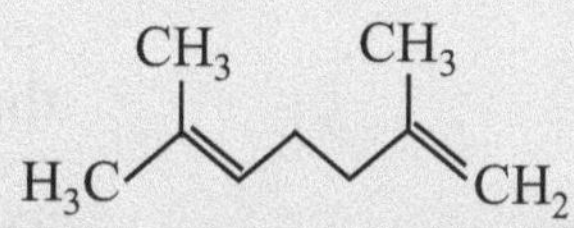
Xanthophylls	Astaxanthin	3,3′-Dihydroxy-β,β-caroteno-4,4′-dione		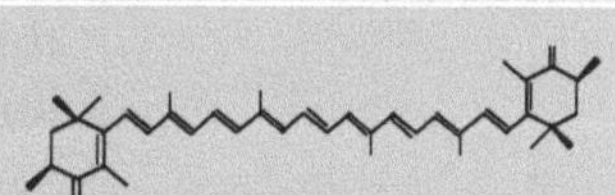
	Capsanthin	3,3′-Dihydroxy-β,κ-carotene-6′-one		
	Canthaxanthin	β,β-Carotene-4,4′-dione		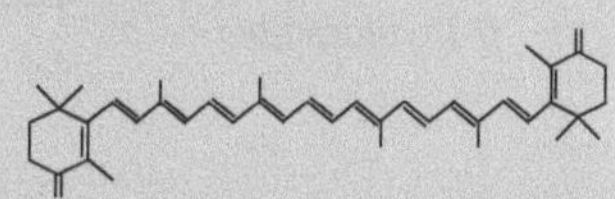
	Lutein	β,ε-Carotene-3,3′-diol	Orange; pumello	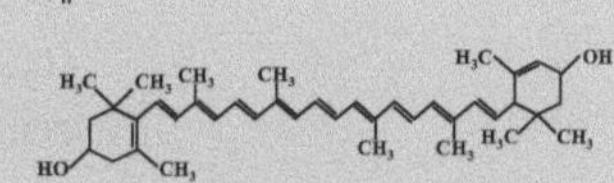
	Zeaxanthin	β,β-Carotene-3,3′-diol	Orange; mandarin	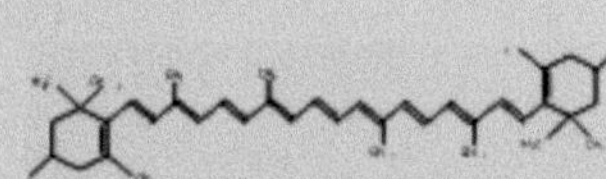
	β-Cryptoxanthin	β,β-Carotene-3-ol	Orange; grapefruit; mandarin; lemon	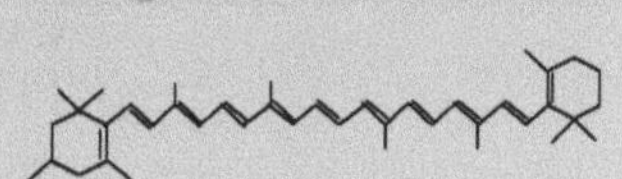

Source: Alquézar et al. [5]; Andrade et al. [1]; Khoo et al. [6]; Kultys and Kurek [7]; Namitha and Negi [8]; Rodrigo et al. [9]; Yahia et al. [3]

oil, rich in lycopene, ζ-carotene, and zeaxanthin, can be used in healthcare products. The by-products produced by citrus wastes can also be utilized as animal or livestock feed due to their rich nutritional composition containing carotenoids, vitamins, enzymes, flavonoids, lipids, and dietary fibers [10]. The use of carotenoids as coloring agents in food products has increased globally. A report stated that the food coloring market was expected to reach 2.34 million dollars in 2025, of which the majority will be of natural components, achieving an economic value of 3.75 million dollars [11]. Carotenoids cannot be synthesized in the body and are vital for human health, making them an essential component to be included in the diet. They also work as provitamin A, a precursor of vitamin A. Vitamin A is derived from carotenoid precursors such as α-carotene, ζ-cryptoxanthin, and ζ-carotene [12]. The functions of vitamin A include regulation of gene expression, improvement in the immune system, prevention of inflammatory diseases, cancer, obesity, and prevention of macular degeneration [13, 14]. The xanthophylls lutein, zeaxanthin, and meso-xanthin helps protect retinal membrane against short-wavelength high-intensity light, thus improving visual acuity [15]. Carotenoids are also used in cosmetics and nutraceuticals due to their UV protection abilities. They also help to improve skin texture. A clinical study revealed that 3–6 mg/day asthaxanthin supplementation for 2–16 weeks improved skin texture, appearance, and moisture content [16].

1.3 Methods for the Determination of Different Carotenoids

1.3.1 β-Cryptoxanthin

There are various ways to determine β-cryptoxanthin, each providing different advantages and applications. The primary methods of determination are HPLC and the spectrophotometric method, coupled with other techniques, such as UV or PDA detection, and solvent extraction.

The HPLC method for determining ζ-crytoxanthin is the most widely used technique. This method offers high sensitivity, specificity, reproducibility, and good recovery. This method is a part of the routine analysis of fruits and vegetables to determine and quantify carotenoids. A study was reported for the analysis of α-cryptoxanthin, β-cryptoxanthin, α-carotene, and β-carotene of nine clones of *Pandanus Conoideus* Oil using HPLC, and ζ-crytoxanthin was found in the range of 3.9–29.4 ng/mg [17]. Another method is the use of spectrophotometry. This method is rarely used to detect ζ-crytoxanthin due to its lower specificity for the particular carotenoid. However, it is quick, economical, and used for preliminary screening or quality control.

The coupling techniques include using liquid chromatography with UV or PDA detection. This is generally used to detect a broad spectrum of carotenoids and can be easily integrated with existing laboratory setups. This method has been beneficial for profiling carotenoids in fruits and vegetables and their derivatives. A study detected the presence of ζ-cryptoxanthin in tangerine but not in carrots, spinach, corn, or tomato [18]. Solvent extraction is another technique to determine ζ-crytoxanthin, where solvents extract the carotenoid from the complex for easy detection through HPLC or LC. This method is effective for raw and processed food products. It is generally used to liberate carotenoids from complex food systems for easy detection. A study reported the extraction of ζ-crytoxanthin from fruits like persimmon and Satsuma mandarin to develop functional and health foods [19].

1.3.2 β-Carotene

Various methodologies have been developed till now to determine β-carotene, each having advantages and applications. These methods include HPLC, UV–Vis spectrophotometry, NIR spectroscopy, FTIR, and gas chromatography (GC)/capillary electrochromatography (CEC).

The HPLC method is the most commonly used method for determining carotenoids in food samples. A study was conducted to determine the content of β-carotene in red pepper powder with and without butylated hydroxytoluene using u-HPLC, showing a limit of detection and limit of quantification of 0.28 and 0.62 μg/mL, and 0.85 and 1.89 μg/mL, respectively [20]. UV–Vis spectroscopy is a rapid, low-cost method for ζ-carotene detection, offering good precision. A study was conducted on raw carrot, sweet potato, and supplemented chicken meat nuggets to determine ζ-carotene content using the solvent extraction coupled UV–Vis spectrophotometric method. Four solvents were used, of which acetone showed the best results, providing 67.8–98.8% recovery [21]. NIR spectroscopy is a nondestructive method that provides rapid analysis, and FTIR is a quick and straightforward method that requires minimal preparation. NIR is highly selective and helps reduce matrix interference, whereas FTIR is useful in characterizing food matrices with high β-carotene content. A study comparing UV–Vis spectroscopy, NIR spectroscopy, and FTIR on the determination of β-carotene in pumpkin (*Cucurbita maxima*) flesh, peel, and seed powder revealed that the NIR method exhibited LOD, 3.4 μg/mL; LOQ, 10 μg/mL; and percent recovery, 92–96% but the β-carotene content went undetected in the case of seed [22]. The gas chromatography (GC)/capillary electrochromatography (CEC) method is a very sensitive method specifically used in specialized research. A study on determining β-carotene content was conducted on carrots, comparing HPLC and GC methods. They reported that GC was equally effective in estimating carotenoid content [23].

1.3.3 Zeaxanthin

Various methodologies can be used to determine zeaxanthin content. The most common technique is HPLC, which can be coupled with multiple methods such as ultrasound-assisted extraction and saponification coupled extraction.

A study was conducted to check if HPLC can be a green, efficient, rapid, and reproducible method for quantifying zeaxanthin when combined with food-grade solvents such as ethanol and water. It was found that the proposed methodology was greener, equal to or more efficient, faster, and more reproducible than the comparative methods [24].

When ultrasound-assisted extraction, an eco-friendly method, is coupled with HPLC, it enhances the speed and yield of the process. It is generally used to analyze functional food ingredients. A study on peach palm fruits' peels compared different extraction techniques coupled with HPLC found that ultrasound-assisted extraction provided the highest carotenoid content (67 mg/100 g) [25]. Saponification coupled extraction helps remove interfering lipids/pigments and allows precise quantification. In a study, saponification was done to damage the cell wall of maize samples using different methods such as frozen biomass homogenization, ultrasound, self-cavitation, mortar mechanical homogenization, homogenizer mortar homogenization, and acid hydrolysis (0.1, 1.0, 2.0 M). Acid hydrolysis with 0.1 M concentration was observed to have recovered maximum zeaxanthin, that is, 4.89 μg/mL [26].

1.3.4 Lutein

There are different methods for determining lutein, offering various advantages and applications. These methods include paper spray ionization mass spectroscopy, alkaline (saponification) method, HPLC, and UHPLC-Q–Orbitrap-HRMS.

The paper spray ionization mass spectroscopy is rapid, precise, and cost-effective in lutein detection in complex food matrices. A study was conducted to analyze lutein content in different fruits and vegetables, and the concentration of lutein was found to be 2.76 ± 0.11 μg/g in green beans, being the lowest, and 51.5 ± 0.05 μg/g in mint stem, being the highest [27]. They stated that the referred technology was used to reduce the duration and complexity of the sample preparation analysis. Another method, the alkaline (saponification) method, and HPLC, is alkaline hydrolysis followed by solvent extraction and analysis using HPLC. This was one of the older techniques used to analyze lutein content in cereals. This method presented high lutein content in green vegetables like spinach and broccoli [28]. A UHPLC-Q–Orbitrap-HRMS method was developed for the analysis of lutein content in marigold flowers, along with the extraction methods, including ultrasonication and microwave-assisted extraction, providing 27.22 mg/g of lutein content [29].

1.3.5 ζ-Carotene

The primary methods used for the determination of ζ-carotene are high-pressure thin layer chromatography (HPTLC) and the use of metabolomics tools. HPTLC is an effective method offering proper separation and quantification of carotenoids, including ζ-carotene, with higher sensitivity and reproducibility [30]. The advancements in instrumentation have made it possible to discover different metabolites in carotenoids with the help of metabolomics tools. They offer three approaches, including targeted analysis, profiling, and fingerprinting. Metabolomics fingerprinting explores different metabolic patterns that vary based on certain applied conditions. Metabolic profiling analyzes the previously chosen set of metabolites of interest. Targeted analysis is particularly for determining a specific known carotenoid and evaluating its profile by varying settings or grouping samples. Majorly, MS and NMR are the two platforms used for metabolomics studies, either alone or in conjunction [31].

1.3.6 Violaxanthin

Violaxanthin, a significant xanthophyll pigment, can be quantified and identified using different analytical methods such as HPLC, UPLC–MS/ LC–MS/MS, spectrophotometric quantification (UV–Vis spectroscopy), gas chromatography (GC), and mass spectrometry (MS).

HPLC is the standard method for determining violaxanthin, which offers high sensitivity and selectivity. This method is mainly used for plant, algae, and food-based samples to monitor violaxanthin content and cycle pigment profiles [32]. Ultra-performance liquid-chromatography–mass spectroscopy (UPLC–MS) is desirable for very complex systems, combining rapid separation with mass detection, allowing both qualitative and quantitative assessment using molecular ion tracing and fragment patterns. It is used in environmental, nutritional, and food quality studies for detection at lower concentrations in mixtures [33]. Spectrophotometric quantification is generally used to quantify violaxanthin for high-throughput or on-site estimation in plants and algal research [34]. Gas chromatography (GC) is rarely used to detect due to its low sensitivity. It is generally used for specialized research [35]. Mass spectrometry (MS) is usually used for molecular weight-based compound detection confirmation. This method does not quantify highly sensitive samples [36].

2 Materials

2.1 β-Cryptoxanthin

The materials required for ζ-cryptoxanthin estimation include β-cryptoxanthin standard, methanol (HPLC grade), acetone, acetonitrile (HPLC grade), appropriate LC column (C18 or C30), chloroform, ascorbyl palmitate, pyrogallol, potassium hydroxide (KOH), sodium chloride, sodium sulfate, n-hexane, ethyl acetate, petroleum ether, ethanol, 2-propanol, and antioxidants: ascorbyl palmitate, pyrogallol, PTFE membrane filters (0.22 μm), C18 HPLC column, UV–Vis spectrophotometer, volumetric flasks, pipettes, and grinding equipment.

2.2 *β*-Carotene

The materials required for β-carotene estimation are β-carotene standard, solvents: ethanol, methanol, acetone, dichloromethane, acetonitrile, methyl tert-butyl ether (MTBE), petroleum ether, antioxidant: butylated hydroxytoluene (BHT), GC or CEC system, β-carotene derivatives standard (requires derivatization, e.g., methylation), derivatization reagents (e.g., BSTFA), PTFE or millipore membrane filters (0.45 μm), C18 or C30 HPLC column, spectrophotometer, and volumetric flasks, pipettes, and rotor homogenizer.

2.3 Zeaxanthin

The materials required for zeaxanthin estimation are zeaxanthin standard (analytical grade), extraction solvents: methanol, acetone, ethanol, n-hexane, (all HPLC grade), potassium hydroxide (KOH, 10–20% in methanol), petroleum ether, ethyl ether, dodium sulfate (anhydrous), PTFE membrane filters (0.22–0.45 μm), C18 or C30 HPLC column, ultrasonic bath or probe system, volumetric flasks, sample homogenizer, and analytical balance.

2.4 Lutein

The material required for lutein content determination are filter paper methanol, ethanol, hexane, ether, dichloromethane, sodium chloride, potassium hydroxide, sodium hydroxide, butylated hydroxytoluene, acetonitrile, internal standard solution, sample extract, sample, lutein standard, high purity nitrogen gas, analytical balance, micro-pipettes, scissors, centrifuge tubes, water bath, LC injection, UHPLC system, Q–Orbitrap high-resolution mass spectrometer, reversed-phase UHPLC column, autosampler vials, and filters.

2.5 *ζ*-Carotene

For the determination of ζ-carotene, the material required are samples containing ζ-carotene, cold acetone, hexane and/or petroleum ether, methanol, ethyl acetate, magnesium carbonate or hyflosupercel and sodium chloride/saturated brine, and HPTLC plates: Silica gel 60 F254 or equivalent, standard solution (pure ζ-carotene), glassware (amber test tubes, pipettes, and volumetric flasks), centrifuge, vortex mixer, HPLC scanner/densitometer, and nitrogen or argon gas, formic acid, butylated hydroxytoluene, internal standards, centrifuge and filtration units (0.22 μm filters), autosampler vials, metabolomics instrument or GC–MS, and CTools/software.

2.6 Violaxanthin

For the determination of violaxanthin using HPLC, the requirements are sample, extraction solvents (HPLC grade): ethanol, acetonitrile, methanol, isopropanol, water (often with 0.1% formic acid), antioxidant (butylated hydroxytoulene), KOH, NaOH, hexane, ethyl acetate, deionized water, pure violaxanthin standard, reverse-phase HPLC column (C18 or C30), mobile phases, autosampler vials and mass spectrometer, derivatization agent, GC-grade solvents, GC–MS instrument, syringe filters (0.22–0.45 μm), and amber vials.

3 Methods

3.1 β-Cryptoxanthin

The step-by-step procedure for ζ-cryptoxanthin estimation through the HPLC method is as follows:

- Extract carotenoids using HPLC-grade solvent (e.g., methanol–chloroform 96:4 v/v) and add antioxidants to prevent carotenoid degradation.
- Saponification (If Matrix is Oily): Treat with KOH in ethanol; incubate at controlled temperature (20–40 °C) for 2–8 h.
- Partition into n-hexane (volume ratio around 1:6–1:4); wash the organic phase with sodium chloride solution. Dry extract with sodium sulfate and filter solution through a 0.22 μm PTFE membrane.
- Inject aliquot (10 μL) into HPLC system (C18 column); flow rate 1 mL/min.
- Isocratic or gradient elution (mobile phase: methanol–chloroform or acetone–water).
- Detect at 450–455 nm.

The step-by-step procedure for ζ-cryptoxanthin estimation through the spectrophotometric method is as follows:

- Filter and dilute the extract appropriately.
- Record absorbance at β-cryptoxanthin's maximum wavelength (450–455 nm).
- Compare absorbance/intensity to the standard curve.

3.2 β-Carotene

The step-by-step procedure for the HPLC method is as follows:

- Add ethanol and shake vigorously. Then, add dichloromethane (120 mL) and shake/homogenize thoroughly.
- Let the mixture settle/cool in the dark for 2 h and filter through a 0.45 μm membrane.
- Concentrate the solution if the content is low; dilute if the content is high.
- Inject 10–20 μL into HPLC (C18/C30 column).

The step-by-step procedure required for the UV–Vis spectrophotometric method is as follows:

- Homogenize and weigh the food sample. Extract β-carotene in acetone or ethanol—boil or soak as needed.
- Filter and bring the extract to a known volume.
- Measure absorbance at 453–455 nm.
- Compare the sample absorbance to the standard curve.

The step-by-step procedure for NIR spectroscopy is as follows:

- Prepare the samples for extraction.
- Scan sample in NIR region (1200–1500 nm; typical absorption at 1415 nm).
- Use the calibration curve from standards.

The step-by-step procedure for FTIR is as follows:

- Homogenize or extract the food sample as appropriate. Then, prepare a sample pellet or solution.
- Record absorbance in region containing β-carotene signature bands (e.g., 1512, 1246 cm^{-1}). Compare with standards.

The step-by-step procedure for gas chromatography (GC)/capillary electrochromatography (CEC) is as follows:

- Extract β-carotene from the sample with acetone/hexane. Dry and derivatize as needed.
- Inject the derivatized sample onto the GC/CEC column.
- Detect at appropriate conditions for β-carotene derivatives.

3.3 Zeaxanthin

The step-by-step procedure for HPLC is as follows:

- Prepare zeaxanthin standard stock solutions in methanol and dilute them to the required concentrations. Store solutions at −20 °C and protect from light.
- Wash, dice, or grind the food samples, homogenize, and weigh a precise portion.
- Extract with an appropriate solvent. Vortex mix for about 1–2 min; incubate at 40 °C for 10–20 min, and filter the extract through PTFE membrane (0.22 μm).
- Inject 10–20 μL into HPLC.
- Use optimized conditions: e.g., C18 column at 30 °C, mobile phase acetone/water gradient (see optimized protocol), flow rate 1 mL/min, detection at 450 nm.

3.4 Lutein

The step-by-step procedure for lutein estimation using paper spray ionization mass spectroscopy is as follows [37]:

- Use a suitable solvent to extract lutein from the sample (e.g., methanol or acetone).
- Load a small amount (5–10 μL) of sample extract onto the filter paper tip. Allow to dry, if required. Position the loaded paper in front of the MS inlet.
- Apply 10–20 μL of methanol: water mixture on the back of the paper.

- Apply a high voltage (usually 1–3 kV) to the paper. Ions are created and sent to the mass spectrometer for examination.
- Quantify using mass spectra and compare to standard calibration if quantitative.

The procedure for the UHPLC-Q–Orbitrap-HRMS method is as follows [36]:

- Filter the extract before injecting and prepare the mobile phase.
- Set up a gradient elution (for example, 0–100% methanol over 5 min; refer to the instrument-specific procedure).
- Injection ranges from 2 to 5 μL every run.
- Use Q-Orbitrap in positive/negative ion mode, full scan, or targeted MS/MS (m/z 100–1000).
- Use external or internal standards. Analyze lutein for accurate mass and confirm fragment ions.
- Use the associated software for chromatogram extraction and quantification.

3.5 ζ-Carotene

The procedure for ζ-carotene content determination using the HPTLC method is as follows [38]:

- Weigh a specified homogenized sample (0.05–1 g).
- Extract carotenoids with cold acetone to separate phases, then vortex and centrifuge.
- To eliminate aqueous pollutants, partition carotenoids into hexane or petroleum ether and wash with saturated NaCl.
- Combine extracts and concentrate in a moderate nitrogen stream.
- Using a microliter syringe, apply a measured quantity of the concentrated extract and ζ-carotene standard on the HPTLC plate.
- Develop the plate in a chamber equilibrated with the mobile phase (for example, hexane:acetone 7:3). Allow migration till the desired solvent front (e.g., 8 cm), then remove and dry the plate.
- Examine carotenoid bands using visible or UV light. Scan the produced plate with an HPTLC scanner/densitometer, usually at 440–460 nm. Identify ζ-carotene by comparing its retention factor (Rf) and absorption. To quantify ζ-carotene, compare band intensity/area to the standard calibration curve.

The procedure using metabolomics tools is as follows [39]:

- Weigh and homogenize the tissue accurately, then add BHT as an antioxidant.

- Extract with an appropriate organic solvent (such as methanol: acetone), vortex, and centrifuge.
- Divide into hexane to improve recovery of nonpolar carotenes, if required.
- Dry the extracts under nitrogen and reconstitute in an injection solvent (e.g., methanol or 50:50 methanol:water).
- Pour the solution through a 0.22 μm filter into an autosampler vial.
- Inject the filtered extract into the LC–MS system through a C18 or comparable column. Set the optimal chromatographic separation gradient. To identify ζ-carotene, obtain high-resolution mass spectra, including the precise m/z and retention duration. Get full-scan MS and data-dependent MS/MS for annotation for untargeted metabolomics.
- Use software to process raw data (such as XCMS, compound discoverer, and MetaboAnalyst). Identify ζ-carotene with the spectrum library and standards for m/z and retention time matching. To quantify, integrate the peak area, and normalize to the internal standard or calibration curve.

3.6 Violaxanthin

The procedure for violaxanthin content determination using the HPLC method is as follows:

- Mix the sample in cold ethanol/methanol with BHT. Centrifuge and collect the supernatant.
- To remove chlorophyll, treat it with KOH in ethanol and neutralize it.
- Combine the extract with hexane/ethyl acetate; separate the organic phase.
- Filter the sample using a 0.22 μm syringe filter into HPLC.
- Use isocratic or gradient elution (e.g., acetonitrile: water and ethyl acetate) with a suitable flow rate. Detect at 440–450 nm.
- Compare the violaxanthin peak area/height with those of calibration standards.

The procedure for violaxanthin content determination using the UPLC–MS/LC–MS/MS method is as follows:

- Mix the sample in cold ethanol/methanol with BHT. Centrifuge and collect the supernatant.
- To remove chlorophyll, treat it with KOH in ethanol and neutralize it.
- Combine the extract with hexane/ethyl acetate; separate the organic phase.
- Filter using a 0.22 μm syringe filter.

- Program the UPLC gradient. Set MS to positive ion mode and scan an appropriate m/z (violaxanthin: [M+H]+ ~601).
- Inject the extracted sample.
- Identify violaxanthin using retention time and MS/MS fragmentation. Calculate using standard violaxanthin calibration curves. Confirm ion identity and purity using fragment ions.

The procedure for violaxanthin content determination using the UV–Vis Spectroscopy method is as follows:

- Prepare by pipetting an aliquot of the extract into a quartz cuvette.
- Measure absorbance from 200 to 700 nm, with maxima at 440–445 nm.
- Prepare and read standards under the same conditions.
- Use a standard curve at the maximum absorption wavelength. Modern approaches like machine learning or MCR-ALS can deconvolute overlapping spectra when dealing with pigment mixes.

The procedure for violaxanthin content determination using the gas chromatography (GC) method is as follows:

- Convert violaxanthin into a volatile derivative and inject the derivative into GC or GC–MS.
- Check for the characteristic mass/retention of the violaxanthin derivative.
- If available, compare to the standard; otherwise, estimate using the peak area and known response parameters.

The procedure for violaxanthin content determination using the mass spectrometry (MS) method is as follows:

- To perform mass analysis, dilute the sample in pure solvent and inject it directly or after LC separation.
- Detect [M+H]+ at m/z 601 (or sodium/potassium adducts).
- Confirm identity by assessing characteristic MS/MS fragments.
- Quantify by comparing MS peak intensity to a standard curve or using an internal isotopic standard for accuracy.

3.7 Assessments/Calculations/Formulas

The assessments required for the HPLC method are as follows:

- Preparation of calibration curve:

$$As = aCs + b$$

where a and b are calculated as

$$C = \frac{(A - A0)Cs}{aCs + b}$$

- The formula for calculating the carotenoid content is

$$Xi\ (\text{mg/g}) = \frac{C \times V \times f}{m} \times \left(\frac{1}{1000}\right)$$

where As = peak area of each carotenoid in the standard solution; Cs = concentration of each carotenoid in the standard solution, mg/L; C = concentration of each carotenoid in the sample solution, mg/L; A = peak area of each carotenoid in the sample solution; $A0$ = peak area of each carotenoid in the sample dilution solution; Xi = the content of each carotenoid in the sample, μg/g; V = extract liquid in mL; f = dilution ratio of sample solution; m = mass of sample in g

The assessments required for the spectrophotometric method are as follows:

- Preparation of Standard Curve: Take absorbance at different known concentrations of pure carotenoid and plot a curve. Use the equation of the curve for further calculations.
- The formula for calculating total carotenoid content is as follows:

$$Xi\ (\mu\text{g/g}) = \frac{A \times V \times 10^4}{A_{1\ \text{cm}}^{1\%} \times P}$$

where Xi = carotenoid content; A = absorbance; V = total extract volume; P = sample weight; $A_{1\ \text{cm}}^{1\%}$ = carotenoid extinction coefficient in a suitable solvent.

4 Notes

The notes for carotenoid content determination are as follows:

- Process 1 sample at a time for consistency.
- Saponify if the food sample is fatty/pigmented.
- Conduct blank trials for baseline correction.
- Store extracts and standards at −20 °C to avoid degradation.
- Optimize LC parameters (column temp, flow rate, and detection wavelength) in the HPLC case via single-factor experiments.
- Adjustment of solvent ratios and extraction conditions is key to maximizing recovery.
- Calibration standards must be freshly prepared.

- Consider isomerization during extraction; optimize parameters for minimal change.
- Extract blanks for baseline correction.
- Chromatographic purity should exceed 95% (peak area).

References

1. Andrade MA, Barbosa CH, Shah MA et al (2023) Citrus by-products: valuable source of bioactive compounds for food applications. Antioxidants 12:38. https://doi.org/10.3390/ANTIOX12010038
2. Thiruvalluvan M, Kaur BP, Singh A, Kumari S (2024) Enhancement of the bioavailability of phenolic compounds from fruit and vegetable waste by liposomal nanocarriers. Food Sci Biotechnol 33:307–325. https://doi.org/10.1007/S10068-023-01458-Z/TABLES/4
3. Yahia EM, de Jesús Ornelas-Paz J, Emanuelli T et al (2017) Chemistry, stability, and biological actions of carotenoids. In: Fruit and vegetable phytochemicals: chemistry and human health, 2nd edn. John Wiley & Sons, Ltd, pp 285–345
4. Meléndez-Martínez AJ, Mandić AI, Bantis F et al (2022) A comprehensive review on carotenoids in foods and feeds: status quo, applications, patents, and research needs. Crit Rev Food Sci Nutr 62:1999–2049. https://doi.org/10.1080/10408398.2020.1867959
5. Alquézar B, Rodrigo MJ, Zacarías L (2008) Carotenoid biosynthesis and their regulation in citrus fruits. Tree For. Sci. Biotechnol 2 (1):23–37
6. Khoo HE, Prasad KN, Kong KW, Jiang Y, Ismail A (2011) Carotenoids and their isomers: color pigments in fruits and vegetables. Molecules 16(2):1710–1738. https://doi.org/10.3390/molecules16021710
7. Kultys E, Kurek MA (2022) Green extraction of carotenoids from fruit and vegetable byproducts: A review. Molecules 27(2):518. https://doi.org/10.3390/molecules27020518
8. Namitha KK, Negi PS (2010) Chemistry and biotechnology of carotenoids. Critical Reviews in Food Science and Nutrition 50(8):728–760. https://doi.org/10.1080/10408398.2010.499811
9. Rodrigo MJ, Alquézar B, Alós E, Lado J, Zacarías L (2013) Biochemical bases and molecular regulation of pigmentation in the peel of Citrus fruit. Scientia Horticulturae 163:46–62. https://doi.org/10.1016/j.scienta.2013.08.014
10. Idamokoro EM, Hosu YS (2022) Out-look on worldwide trends of related studies on citrus waste as feed for livestock production: a scientometric analysis. Front Res Metr Anal 7: 869974. https://doi.org/10.3389/frma.2022.869974
11. Li Z, Li C, Cheng P, Yu G (2022) Rhodotorula mucilaginosa—alternative sources of natural carotenoids, lipids, and enzymes for industrial use. Heliyon 8:e11505
12. Gürbüz M, Aktaç Ş (2022) Understanding the role of vitamin A and its precursors in the immune system. Nutr Clin Metab 36:89–98
13. Arias A, Feijoo G, Moreira MT (2022) Exploring the potential of antioxidants from fruits and vegetables and strategies for their recovery. Innov Food Sci Emerg Technol 77:102974. https://doi.org/10.1016/j.ifset.2022.102974
14. Kabir MT, Rahman MH, Shah M et al (2022) Therapeutic promise of carotenoids as antioxidants and anti-inflammatory agents in neurodegenerative disorders. Biomed Pharmacother 146:112610
15. Saini RK, Prasad P, Lokesh V et al (2022) Carotenoids: dietary sources, extraction, encapsulation, bioavailability, and health benefits—a review of recent advancements. Antioxidants 11:795
16. Ng QX, De Deyn MLZQ, Loke W et al (2021) Effects of astaxanthin supplementation on skin health: a systematic review of clinical studies. J Diet Suppl 18:169–182
17. Sarungallo ZL, Hariyadi P, Andarwulan N et al (2015) Analysis of α-cryptoxanthin, β-cryptoxanthin, α-carotene, and β-carotene of pandanus conoideus oil by high-performance liquid chromatography (HPLC). Procedia Food Sci 3:231–243. https://doi.org/10.1016/j.profoo.2015.01.026
18. Konings EJM, Roomans HHS (1997) Evaluation and validation of an LC method for the analysis of carotenoids in vegetables and fruit. Food Chem 59:599–603. https://doi.org/10.1016/S0308-8146(96)00343-3
19. Alimentos E. Nutricao, Rodriguez GA, Bieri JG (2014) Beta-cryptoxanthin from plant source and a process for its preparation. J Agric Food Chem 18:2064–2070

20. Ha J, Seo HY, Shim YS et al (2010) Determination of capsaicinoids in foods using ultra high performance liquid chromatography. Food Sci Biotechnol 19:1005–1009. https://doi.org/10.1007/s10068-010-0141-8
21. Biswas AK, Sahoo J, Chatli MK (2011) A simple UV-Vis spectrophotometric method for determination of β-carotene content in raw carrot, sweet potato and supplemented chicken meat nuggets. LWT 44:1809–1813. https://doi.org/10.1016/j.lwt.2011.03.017
22. Hagos M, Redi-Abshiro M, Chandravanshi BS, Yaya EE (2022) Development of analytical methods for determination of β-carotene in pumpkin (Cucurbita maxima) flesh, peel, and seed powder samples. Int J Anal Chem 2022: 9363692. https://doi.org/10.1155/2022/9363692
23. Mangunsong S, Taswin M, Natarajan SB (2020) Determine β-carotene in carrot (Daucus carota L.) by using HPLC and GC-MS. IOSR J Pharm 10:21–27
24. Carneiro AM, Lima BR, Chibli LA et al (2023) An updated procedure for zeaxanthin and lutein quantification in corn grains based only in water and ethanol. Food Chem 427: 136589. https://doi.org/10.1016/j.foodchem.2023.136589
25. Menezes Silva JV, Silva Santos A, Araujo Pereira G, Campos Chisté R (2023) Ultrasound-assisted extraction using ethanol efficiently extracted carotenoids from peels of peach palm fruits (Bactris gasipaes Kunth) without altering qualitative carotenoid profile. Heliyon 9:e14933. https://doi.org/10.1016/j.heliyon.2023.e14933
26. Butnariu M, Rodino S, Petrache P et al (2014) Determination and quantification of maize zeaxanthin stability. Dig J Nanomater Biostruct 9: 745–755
27. Alsaggaf WT (2022) Rapid determination of lutein in fresh and commercial food samples using paper spray ionization mass spectrometry. LWT 155:112915. https://doi.org/10.1016/j.lwt.2021.112915
28. Fratianni A, Mignogna R, Niro S, Panfili G (2015) Determination of lutein from fruit and vegetables through an alkaline hydrolysis extraction method and HPLC analysis. J Food Sci 80:C2686–C2691. https://doi.org/10.1111/1750-3841.13122
29. Suwanklang P, Thilavech T, Limwikrant W et al (2023) Analysis of lutein content in microencapsulated marigold flower extract (Tagetes erecta L.) using UHPLC-Q-Orbitrap-HRMS and its cytotoxicity in ARPE-19 cells. Molecules 28:6025. https://doi.org/10.3390/MOLECULES28166025
30. Alqarni MH, Alam P, Alam A et al (2021) A greener hptlc approach for the determination of β-carotene in traditional and ultrasound-based extracts of different fractions of daucus carota (L.), ipomea batatas (l.), and commercial formulation. Agronomy 11:2443. https://doi.org/10.3390/agronomy11122443
31. Amorim-Carrilho KT, Cepeda A, Fente C, Regal P (2014) Review of methods for analysis of carotenoids. TrAC Trends Anal Chem 56: 49–73
32. Ahmad N, Mounsef JR, Lteif R (2021) A simple and fast experimental protocol for the extraction of xanthophylls from microalga Chlorella luteoviridis. Prep Biochem Biotechnol 51:1071–1075. https://doi.org/10.1080/10826068.2021.1901231
33. Fu W, Magnúsdóttir M, Brynjólfson S et al (2012) UPLC-UV-MSE analysis for quantification and identification of major carotenoid and chlorophyll species in algae. Anal Bioanal Chem 404:3145–3154. https://doi.org/10.1007/s00216-012-6434-4
34. Victor P, Camarenaa-Bernard C (2022) Lutein, violaxanthin, and zeaxanthin spectrophotometric quantification: a machine learning approach. J Appl Phycol 35:73–84
35. Abrahamsson V, Rodriguez-Meizoso I, Turner C (2012) Determination of carotenoids in microalgae using supercritical fluid extraction and chromatography. J Chromatogr A 1250: 63–68. https://doi.org/10.1016/j.chroma.2012.05.069
36. Wang P, Xu Q, Zou H et al (2024) Development and application of UPLC-Q-Orbitrap HRMS methods for the determination of eight metabolites of p-chloronitrobenzene in human urine. J Chromatogr B Anal Technol Biomed Life Sci 1244:124254. https://doi.org/10.1016/j.jchromb.2024.124254
37. Bao Y, Guo N, Hu X et al (2024) Chemical derivatization and paper spray ionization mass spectrometry for fast screening of retinoic acid in cosmetics. Molecules 29:4491. https://doi.org/10.3390/molecules29184491
38. Zeb A, Murkovic M (2010) Thin-layer chromatographic analysis of carotenoids in plant and animal samples. JPC-J Planar Chromat 23:94–103
39. Liu C, Xu Q, Liu Y et al (2024) Metabolomic analysis of carotenoids biosynthesis by Sphingopyxis sp. USTB-05. Molecules 29:4235. https://doi.org/10.3390/molecules29174235

Chapter 8

Pectin Quantification in Citrus Peels

Priyanka Dubey, Rahul Singh, Alvina Farooqui, and Owais Yousuf

Abstract

Citrus peel waste, a major by-product of citrus fruit processing, is an abundant source of pectin, a polysaccharide with significant industrial value due to its gelling, stabilizing, and emulsifying properties. This chapter explores methodologies for extracting, quantifying, and characterizing pectin, focusing on sustainability and maximizing economic efficiency. Various approaches are discussed in detail, including gravimetric, titrimetric, colorimetric, and spectrophotometric methods. Advanced techniques like FTIR and HPLC provide precise insights into pectin's structural and functional attributes. Protocols such as alcohol precipitation, calcium pectate precipitation, and carbazole assays are evaluated for their accuracy, efficiency, and reproducibility. By comparing the strengths and limitations of different methods, this chapter provides a valuable resource for researchers and industry professionals to estimate pectin in a constructive and precise manner. It emphasizes the potential of citrus peel waste as a sustainable, high-value ingredient, contributing to environmental conservation and economic sustainability.

Key words Citrus peel, Pectin, Extraction, Characterization, Sustainability, FTIR, HPLC

1 Introduction

Citrus peel waste, an abundant by-product of the global citrus processing industry, has emerged as a significant resource for pectin, a structural heteropolysaccharide primarily found in the primary cell walls and intercellular regions of higher plants [1]. This polysaccharide is predominantly composed of α-galacturonic acid units (at least 65%) with varying levels of methyl esterification [2]. In recent years, the industrial extraction of pectin has shifted towards utilizing fruit peels, particularly from citrus fruits, as a primary source processed annually within the juicing industry. This is largely due to the vast quantities of citrus fruits such as lemons, oranges, grapefruit, and sweet lime, which produce millions of tons of peel as a by-product, constituting nearly half the fruit's weight, and are often discarded or underutilized, contributing to environmental concerns and economic inefficiencies [3]. Pectin imparts valuable functional properties such as gelling,

Shweta Suri et al. (eds.), *Citrus Fruit Processing By-Products: Technologies and Basic Protocols*, Methods and Protocols in Food Science, https://doi.org/10.1007/978-1-0716-5088-2_8,

stabilizing, and emulsifying capabilities. As the demand for natural and sustainable ingredients grows, accurate determination of pectin content in citrus peel waste has become essential for maximizing its industrial potential. Pectin quantification involves evaluating its physicochemical properties, such as solubility, molecular weight, degree of esterification, and galacturonic acid content, to determine its quality and suitability for specific applications [4]. Conventional gravimetric methods, which rely on extracting, drying, and weighing pectin residues, have long been the standard but are labor-intensive and time-consuming [5]. To address these limitations, modern analytical techniques have been developed, including spectrophotometric methods like carbazole reagent assays and the Xylenol orange method, which provide rapid assessments of pectin concentration through colorimetric reactions [6, 7]. Chromatographic methods, such as high-performance liquid chromatography (HPLC), offer detailed insights into the molecular structure of pectin, enabling precise measurements of esterification levels and structural heterogeneity [8]. Additionally, advanced spectroscopic techniques, including Fourier-transform infrared (FTIR) spectroscopy, allow for nondestructive analysis of pectin's functional groups, chemical bonds, and monosaccharide composition, enhancing the understanding of its structural and functional attributes [9]. Complementing this, an advanced copper ion exchange method has also been used. The choice of method depends on factors such as required sensitivity, accuracy, and the complexity of the citrus peel matrix, often necessitating a combination of approaches to achieve comprehensive results [10]. Accurate and reproducible quantification not only facilitates the optimization of extraction protocols but also ensures consistency in pectin quality for industrial use. This growing focus on the valorization of citrus peel waste through improved pectin determination techniques underscores its potential to transform agricultural by-products into high-value ingredients, promoting sustainability and economic efficiency in the citrus industry. In this chapter, various protocols for determining pectin in citrus peels, focusing on the advantages and limitations of each method, have been discussed in detail. This chapter aims to provide a comprehensive resource for researchers and industry professionals seeking to develop efficient, accurate, and reproducible methods for the quantification of pectin in citrus peels. By highlighting the strengths and weaknesses of existing techniques, this chapter will guide future research in pectin extraction and characterization, facilitating the more sustainable utilization of citrus peels as a valuable natural resource.

2 Materials

Always use purified deionized water and analytical-grade reagents for all solution preparations. Reagents should be freshly prepared and stored at the specified temperature for each reagent.

2.1 Calcium Pectate Gravimetric Method

1. For 1 N acetic acid, dilute 30 mL of glacial acetic acid to 500 mL with distilled water.
2. For 1 N calcium chloride solution, dissolve 27.5 g of anhydrous calcium chloride ($CaCl_2$) in water and dilute the solution to a final volume of 500 mL.
3. To prepare 1% silver nitrate solution, dissolve 1 g of silver nitrate ($AgNO_3$) in 100 mL of distilled water.
4. For 0.01 N hydrochloric acid (HCl), dilute 0.85 mL of concentrated HCl (37%, approximately 12 N) to 1 L with distilled water.
5. For 0.05 N hydrochloric acid, dilute 4.25 mL of concentrated HCl to 1 L with distilled water.
6. Lastly, for 0.3 N hydrochloric acid, dilute 25 mL of concentrated HCl to 1 L with distilled water.

2.2 Carbazole Spectrophotometric Assay

1. To prepare 60% ethyl alcohol, simply mix 500 mL of 95% ethyl alcohol with 300 mL of water.
2. The 95% ethyl alcohol can be sourced commercially.
3. If purified ethyl alcohol is required, it can be made by refluxing 1 L of 95% ethyl alcohol with 4 g of zinc dust and 2 mL of concentrated sulfuric acid (H_2SO_4) for 15 h, followed by distillation using a glass apparatus. To refine it further, redistill with 4 g of zinc dust and 4 g of potassium hydroxide (KOH).
4. For the sodium hydroxide solutions, prepare standardized solutions of 1 N and 0.05 N concentrations as needed.
5. The 0.1% carbazole reagent is made by dissolving 100 mg of recrystallized carbazole in purified ethyl alcohol and diluting it to a total of 100 mL.
6. Additionally, analytical grade sulfuric acid (H_2SO_4) is required for specific steps in the process.

2.3 Copper Ion Colorimetry Method

1. Cupric Sulfate Solution: Dissolve 25 g of cupric sulfate ($CuSO_4$) in 1 L of distilled water to prepare a 1% Cu^{2+} solution.
2. Phenol Solution: Prepare a 5% phenol solution by dissolving 5 g of phenol in 100 mL of distilled water (*see* **Note 6**).
3. Sulfuric Acid Solution: Dilute concentrated sulfuric acid (H_2SO_4) with distilled water to prepare a 50% sulfuric acid solution.

2.4 Xylenol Orange Dye Method

1. Xylenol Orange Dye Solution: Dissolve 0.1 g of xylenol orange in 100 mL of distilled water to make a 0.1% (w/v) solution. Store in a dark container to avoid degradation by light. The solution should be freshly prepared or stored for short durations in a refrigerator (*see* **Note 6**).
2. Sodium Acetate Buffer (pH 4.5): Dissolve 6.9 g of sodium acetate trihydrate in 900 mL of distilled water, adjust pH to 4.5 with acetic acid, and dilute to 1 L.
3. Hydrochloric Acid (HCl): Prepare a 0.1 M HCl solution by diluting 8.5 mL of concentrated HCl in distilled water to 1 L. Store in a labeled container and use it to adjust the pH of the extraction process or reaction.
4. Ethanol (95%): Use 95% ethanol to wash the peel or precipitate soluble substances during extraction. Due to its flammable nature, store it in a well-labeled container away from heat.

2.5 Acid–Base Titrimetric Methods

1. To prepare 0.1 M sodium hydroxide (NaOH) solution, dissolve 4 g of NaOH pellets in 1 L of deionized water. Ensure the solution is well-mixed and store it in a sealed container. Next, prepare for the preparation of 0.25 M sodium hydroxide (NaOH) solution, dissolve 10 g of NaOH pellets in 1 L of deionized water, mixing thoroughly to achieve complete dissolution, and store it in a sealed container (*see* **Note 8**).
2. For the 0.25 M hydrochloric acid (HCl) solution, dilute 21.4 mL of concentrated HCl (37%) to 1 L with deionized water, mix thoroughly, and store the solution in a sealed container.
3. To prepare the phenol red indicator by dissolving 100 mg of phenol red in 100 mL of 95% ethanol (*see* **Note 6**).

2.6 Infrared Spectroscopy (FTIR)

1. Use spectroscopic-grade KBr powder, which should be free from water and impurities (*see* **Note 10**).
2. Use dried pectin extracted from citrus peel as described in the HPLC protocol (*see* **Note 9**).

2.7 High-Performance Liquid Chromatography (HPLC)

1. HPLC setup (*see* **Note 5**).
2. For 95% ethanol, mix 95 mL of absolute ethanol with 5 mL of distilled water to prepare 100 mL of 95% ethanol.
3. For 0.1 M HCl—Dilute 8.33 mL of concentrated HCl in about 900 mL of distilled water, then make up the volume to 1 L (*see* **Note 6**).
4. Acetone—Typically used in a *5–50%* concentration in water or aqueous buffers as a mobile phase modifier.

5. For 0.1 N NaOH—Weigh 4 g of NaOH pellets and dissolve in 900 mL of distilled water. Stir until completely dissolved and make up the volume to 1 L.
6. For HPLC-grade water, use a water purification system that produces ultrapure water (resistivity of ≥18.2 MΩ·cm).
7. Pre-purified HPLC-grade methanol or acetonitrile.

3 Methods

3.1 Calcium Pectate Gravimetric Method

1. Weigh 50 g of the blended sample into a 1 L beaker and add 300 mL of 0.01 N HCl.
2. Boil the mixture for 30 min, then filter under suction. Wash the residue with hot water and collect the filtrate.
3. Add 100 mL of 0.05 N HCl to the residue, boil for 20 min, filter, wash, and collect the filtrate.
4. Add 100 mL of 0.3 N HCl to the residue, boil for 10 min, filter, wash, and collect the filtrate.
5. Combine all filtrates, cool the solution and dilute it to a final volume of 500 mL.
6. Take 100–200 mL aliquots from the combined solution and transfer them into 1 L beakers.
7. Add 250 mL of distilled water and neutralize the solution using 1 N NaOH with phenolphthalein as the indicator.
8. After neutralization, add an excess of 10 mL 1 N NaOH while stirring constantly, then let the solution stand overnight.
9. Add 50 mL of 1 N acetic acid, stir and wait for 5 min. Then, add 25 mL of 1 N calcium chloride solution with constant stirring and let the mixture stand for 1 h.
10. Boil the solution gently for 1–2 min.
11. Filter the mixture through pre-weighed Whatman No. 1 filter paper (*see* **Note 1**).
12. Wash the precipitate with near-boiling water until the filtrate test gets negative for chloride using silver nitrate.
13. Transfer the filter paper containing calcium pectate to a weighing dish, dry it overnight at 100 °C, cool it in a desiccator, and weigh it.
14. The pectin content is expressed as % calcium pectate:

$$\%\text{Calcium pectate} = \text{Wt. of calcium pectate} \times 500 \times 100 \div \text{mL of filtrate taken} \times \text{Wt. of sample for estimation}$$

3.2 Carbazole Spectrophotometric Assay

3.2.1 Preparation of Standard Curve

1. Prepare a series of pectin solutions with known concentrations (e.g., 10, 20, 30, 40, and 50 μg/mL) by dissolving standard pectin in distilled water.
2. Add 6 mL of concentrated sulfuric acid to 2 mL of each standard solution and cool to room temperature.
3. Add 0.2 mL of 0.1% carbazole solution and mix thoroughly.
4. Incubate the mixture at 25 °C for 30 min to develop color.
5. Measure the absorbance at 530 nm using a UV–Vis spectrophotometer.

3.2.2 Analysis of Extracted Pectin

1. Dissolve 10 mg of the extracted pectin in 10 mL of distilled water.
2. Follow the same procedure as for the standard solutions (add sulfuric acid, carbazole, and incubate) (*see* **Note 2**).
3. Measure the absorbance of the sample solution at 530 nm.
4. Use the standard curve to determine the concentration of pectin in the sample solution.
5. Record the absorbance of standard and sample solutions.
6. Plot the standard curve and calculate the pectin concentration of the sample.
7. Calculate the total pectin content in the citrus peel using the formula:

$$\text{Pectin content (mg/g)} = \text{Concentration (mg/mL)} \times \text{Volume (mL)}/(\text{Weight of peel sample (g)})$$

3.3 Copper Ion Colorimetry Method

1. For establishment of the Cu^{2+} working curve, 02 mL of different concentrations (3~10.5 mmol/L) of $CuCl_2$ solution were put into a set of 10 mL of the volumetric flask containing buffer solution, respectively [7] (*see* **Note 3**).
2. The resulting solution is transferred to the quartz cuvette, and the buffer solution is used as a blank sample.
3. The absorbance at 712 nm is determined by a UV–Vis spectrophotometer.
4. A linear standard curve is established between the absorbance and the content of Cu^{2+}.
5. For determination of Pectin content, 02 mL of filtrate in calcium pectate was added into 10 mL of the volumetric flask, which contains 8 mL of buffer solution.
6. After mixing well, the mixture is moved to a quartz cuvette and the absorbance at 712 nm is measured in a UV–Vis spectrophotometer to determine the content of Cu^{2+} in the filtrate.

7. The content of Cu^{2+} is further used to calculate the pectin content. The calculation formula is as follows:

$$X = (n_0 - C \times V \times 10^{-3}) \times 64/0.1172 \times m_0 \times (100 - w) \times 10^3 \times 100\%$$

where X is pectin content (measured by copper pectate content, in %); C is the concentration of Cu^{2+} according to the working curve (in mmol/L); n_0 is the content of Cu^{2+} before the reaction (in mmol); V is the total volume of the filtrate (in mL); 64 is the relative atomic weight of copper; 0.1172 is the ratio of copper atoms in copper pectate; m_0 is the weight of the sample (in g); and w is the moisture of the sample (in %).

3.4 Xylenol Orange Dye Method

1. Weigh approximately 5 g of dried citrus peel (or fresh peel if applicable) using an analytical balance.
2. Homogenize the peel in 50 mL of distilled water using a blender or mortar and pestle.
3. Filter the homogenized sample through cheesecloth or filter paper to obtain the extract.
4. Centrifuge the filtered sample at 3000 rpm for 10 min to remove any remaining solid particles. Collect the supernatant for analysis.
5. Prepare standard pectin solutions of known concentrations by dissolving pectin (commercially available) in distilled water.
6. Dilute the stock solution with distilled water to create a series of pectin solutions (e.g., 0.1, 0.2, 0.3, 0.4, and 0.5 mg/mL).
7. Each concentration is used to create a standard curve for pectin quantification.
8. Add 1 mL of the citrus peel extract or standard pectin solution to each test tube or cuvette.
9. Add 1 mL of xylenol orange dye solution to each tube. The dye solution should be prepared by dissolving xylenol orange in an appropriate solvent (usually distilled water) at a concentration of about 0.1% (w/v) (*see* **Notes 4** and **6**).
10. Add 1 mL of sodium acetate buffer (pH 4.5) to each tube. Mix thoroughly.
11. Incubate the mixtures at room temperature for 15–20 min to allow the reaction to proceed.
12. After incubation, measure the absorbance of each solution at 570 nm using a spectrophotometer (*see* **Note 5**).
13. Prepare a blank using the buffer and solvent without any pectin or extract to calibrate the spectrophotometer.

14. Plot a standard curve using the absorbance values obtained from the standard pectin solutions against their respective concentrations.
15. The pectin concentration of the citrus peel extract can be determined by comparing its absorbance value to the standard curve.
16. From the standard curve, determine the concentration of pectin in the citrus peel extract.
17. The equation to calculate the total pectin content in the original citrus peel sample is as follows:

$$\text{Pectin (mg)} = C \times V_{\text{extract}}$$

where C = concentration of pectin in the extract (mg/mL); V_{extract} = volume of extract used for the test (mL).

3.5 Acid–Base Titrimetric Methods

1. 0.5 g of pectin, 5 mL of ethanol (96%), 1 g of sodium chloride, and 100 mL of deionized water were added.
2. The mixture was stirred until the pectin was fully dissolved.
3. The mixture is titrated with 0.1 M NaOH with phenol red as an indicator (*see* **Note 7**).
4. Then, 25 mL of 0.25 M NaOH was added while stirring, to de-esterify pectin.
5. After 30 min 25 mL of 0.25 M HCl was added to neutralize NaOH.
6. The solution was titrated again with 0.1 M NaOH, using phenol red as an indicator, until the color changed (*see* **Note 8**).

3.6 Infrared Spectroscopy

1. For FTIR sample preparation, mix 1 mg of the dried pectin powder with 100 mg of potassium bromide (KBr) powder (*see* **Note 9**).
2. Grind the mixture thoroughly in an agate mortar to ensure uniform mixing. Compress the mixture into a thin pellet using a hydraulic press under vacuum (*see* **Note 10**).
3. After the pellet formation, calibrate the FTIR spectrophotometer using a background scan (e.g., KBr blank for pellet method or air for ATR).
4. Place the prepared KBr pellet or pectin powder (ATR) into the sample chamber (*see* **Note 11**).
5. Record the FTIR spectrum in the range of 4000–400 cm^{-1} with a resolution of 4 cm^{-1}.
6. For data analysis, analyse the characteristic absorption peaks in the FTIR spectrum of pectin at 3390.6, 2939.0, 1749.0, and 1052.1 cm^{-1}, which are related to –OH (hydroxyl groups), –CH (aliphatic chains), C=O (esterified carboxylic acid groups)

group of ester and acid, and –COC– stretching of the galacturonic acid. In the fingerprint region (1200–900 cm^{-1}) of pectin spectra, bands at 1140 and 1040 cm^{-1} were ascribed to stretching vibrations of glycosidic bonds (C–O) and pyranoid rings (C–C).

7. Compare the sample spectrum with a standard pectin FTIR spectrum to confirm the presence of pectin.
8. To quantify the DE of pectin, a calibration curve based on the pectin standard (known as DE) can be established by using the formula:

$$A_{17450}/(A_{1745} + A_{1630}) \times 100$$

where A_{1630} and A_{1745} denote the absorbance intensity, respectively, at 1630 and 1745 cm^{-1} for non-methyl-esterified carboxyl groups and the methyl-esterified carboxyl groups, respectively.

3.7 High-Performance Liquid Chromatography (HPLC)

1. Dissolve 10 mg dried pectin in 10 mL of 0.1 N NaOH (*see* **Note 13**).
2. Heat at 90 °C for 30 min to release galacturonic acid. Neutralize with 0.1 N sulfuric acid. Filter solution through a 0.45 μm membrane for HPLC analysis.
3. To establish HPLC setup (*see* **Note 12**).
4. Prepare standards using galacturonic acid solutions (10–100 μg/mL).
5. To prepare Mobile Phase, use 0.005 M sulfuric acid or phosphate buffer (pH 2–3).
6. Inject 20 μL of standards and sample solutions and record retention times and peak areas.
7. Plot calibration curve (peak area vs. concentration for standards).
8. To quantify pectin content, compare the sample's retention time and peak area with the standard calibration curve.
9. Compare the sample peak area to the calibration curve to determine galacturonic acid concentration.
10. Calculate pectin content using the formula:

$$\text{Pectin content } (\%) = (\text{Galacturonic acid concentration} \times \text{Total volume})/\text{Sample weight} \times 100$$

4 Notes

1. To prepare the filter paper, follow these steps:
   - Wet the filter paper thoroughly with hot water.
   - Then dry it in an oven at 102 °C for 2 h.
   - Once dried, allow the filter paper to cool in a desiccator.
   - Finally, weigh it in a covered dish to ensure accurate measurement.
2. Be cautious with sulfuric acid and carbazole, ensuring proper mixing and color development to avoid errors in absorbance readings.
3. Avoid contamination of the sample by ensuring clean cuvettes, and handle Cu^{2+} solutions with care as they can be reactive.
4. The sodium acetate buffer (pH 4.5) must be prepared with precision, ensuring the pH is accurately adjusted using either acetic acid or sodium acetate to maintain optimal conditions for the dye–pectin interaction.
5. Ensure that all glassware used for reagent preparation is clean and free from contaminants, as impurities may interfere with the reaction and subsequent spectrophotometric analysis.
6. This dye should be stored in an amber bottle to protect it from light.
7. The phenol red indicator must be used carefully, and the titration endpoint should be identified precisely to avoid over- or under-titration, which could lead to inaccurate pectin content calculations.
8. Ensure that all glassware is clean and free from residues that could affect the titration, and perform titrations in a controlled environment to maintain consistency and precision throughout the procedure.
9. Handle KBr pellets gently to avoid breakage, and ensure uniform sample dispersion for consistent results in FTIR analysis.
10. During pellet formation, the mixture should be compressed under vacuum to avoid air bubbles or uneven distribution, which could result in poor-quality spectra.
11. Before analysis, calibrate the FTIR instrument with a background scan (using KBr or air) to eliminate any baseline noise or interference.
12. HPLC Setup should be as follows—Column: Use a C18 column or a carbohydrate-specific column, depending on the analyte. Mobile Phase: Isocratic elution with water and methanol or acetonitrile (e.g., 80:20 v/v). Flow Rate: Set at 1.0 mL/

min, Injection Volume: Typically, 20 μL. Detection: UV detection at 210–220 nm (sensitive to pectin or galacturonic acid).

13. If using extracted pectin, dry it at 40–50 °C until a constant weight is achieved.
14. Handle chemicals carefully, following safety guidelines to avoid contamination or hazards during preparation. Dispose of all waste responsibly, adhering to relevant environmental regulations and safety protocols.

References

1. Yi L, Cheng L, Yang Q, Shi K, Han F, Luo W, Duan S (2024) Source, extraction, properties, and multifunctional applications of pectin: a short review. Polymers 16(20):2883
2. Grassino AN, Barba FJ, Brnčić M, Lorenzo JM, Lucini L, Brnčić SR (2018) Analytical tools used for the identification and quantification of pectin extracted from plant food matrices, wastes, and by-products: a review. Food Chem 266:47–55
3. Dubey P, Tripathi G, Mir SS, Yousuf O (2023) Current scenario and global perspectives of citrus fruit waste as a valuable resource for the development of food packaging film. Trends Food Sci Technol 141:104190
4. Naqash F, Masoodi FA, Rather SA, Wani SM, Gani A (2017) Emerging concepts in the nutraceutical and functional properties of pectin—a review. Carbohydr Polym 168:227–239
5. Pojić M, Kravić S, Stojanović Z, Nollet LML, Toldra F (2015) Analytical methods for determination of moisture and ash in foodstuffs. In: Handbook of food analysis, vol 1. CRC Press, pp 275–296
6. Kurita O, Fujiwara T, Yamazaki E (2008) Characterization of the pectin extracted from citrus peel in the presence of citric acid. Carbohydr Polym 74(3):725–730
7. Wang F, Du C, Chen J, Shi L, Li H (2021) A new method for determination of pectin content using spectrophotometry. Polymers 13(17):2847
8. Walter RH (1991) Analytical and graphical methods for pectin. In: The chemistry and technology of pectin. Elsevier, pp 189–225
9. Brummer Y, Cui SW (2006) Detection and determination of polysaccharides in foods. In: Food polysaccharides and their applications. Taylor & Francis, Boca Raton, p 675
10. Ranganna S (1977) Manual of analysis of fruit and vegetable products. Tata McGraw-Hill, New Delhi

Chapter 9

Separation and Identification of Organic Acids in Citrus By-Products

Owais Yousuf, Seerat Manzoor, Jan Sabura Shabeer, and Kaiser Younis

Abstract

The separation and identification of organic acids in citrus by-products are critical for evaluating their nutritional and industrial potential. Various techniques focusing on precision, efficiency, and scalability have been developed to achieve this. A high-performance liquid chromatographic (HPLC) separation method with photo-diode array detection has been developed for the simultaneous determination of organic acids and phenolic compounds in juices and drinks. Gas chromatography (GC), often combined with mass spectrometry (GC-MS), provides enhanced identification capabilities for volatile organic acids. Ion chromatography (IC) is another reliable method for quantifying organic acids in citrus by-products. Pre-treatment processes of filtration, centrifugation, and derivatization are critical to improve the accuracy of these techniques. These methods enable the characterization of organic acids for applications in food, pharmaceuticals, and agricultural industries, contributing to the sustainable use of citrus processing residues.

Key words Organic acids, Phytochemicals, High-performance liquid chromatography, Flavedo, Albedo, Phenolic compounds

1 Introduction

Citrus Fruits belong to the genus *Citrus L.*, subfamily *Aurantioideae* in the family *Rutaceae*, represent the major fruit crops commercially cultivated worldwide and consist of different forms such as trees, shrubs, and herbs in the world. Citrus fruits are widely known for their health benefits and are the most widely consumed fruits globally. The cultivation of the *Citrus* genus includes the species such as lemon (*C. limon* (L.) Osbeck), sweet orange (*C. sinensis* (L.) Osbeck), mandarin (*C. reticulata* Blanco), grapefruit (*C. paradisi* Macfad.), pomelo (*C. maxima* (Burm.) Merr.), citron (*C. medica* L.), lime (*C. aurantiifolia* (Christm.) Swingle), and bergamot (*C.* × *bergamia* Risso & Poiteau) [1]. Even though a significant part of the industry processes Citrus fruits to create value-added products like juices, concentrates, essential oils, citrus-based candies, etc. However, the by-products produced

Shweta Suri et al. (eds.), *Citrus Fruit Processing By-Products: Technologies and Basic Protocols*, Methods and Protocols in Food Science, https://doi.org/10.1007/978-1-0716-5088-2_9,

yearly have become an environmental as well as economic concern. It contributes to Water pollution, Odor pollution, Methane emissions, Soil imbalance, Pest Attraction, Energy loss, Regulatory fines, Compliance costs, etc.

Waste Valorization can be an effective solution to address these environmental and economic issues as citrus by-products are considered a valuable source of phytochemicals (such as *d*-limonene, essential oils, phenolic acids, carotenoids, vitamins, minerals, and flavonoids) whether, isolated or in the form of mixtures/extracts, could exhibit versatile biological activities especially beneficial for the food industry [1]. The waste generated from citrus fruits includes Peels (almost 50%), Seeds (40%), Pomace, and Wastewater. The Outer layer of Citrus fruit consists of a peel which is divided into two regions: flavedo and albedo. The Flavedo region (epicarp) is bright colored (yellow, orange, or green) and rich in Pigments (carotenoids and chlorophyll), essential oils, and flavonoids. Albedo region (mesocarp) is the white, spongy inner layer of peel located between the flavedo and the fruit segment. It is rich in dietary fiber, flavonoids, phenolic compounds, and some vitamin C. It is low in essential oil compared to flavedo. Both flavedo and albedo of citrus fruit contain organic acids such as citric acid, phenolic acid, ascorbic acid, and malic acid that provide a tangy, citrusy flavor and act as natural preservatives.

The full potential of citrus seed utilization is still an area of interest for the scientific community, as there have been fewer studies dealing with citrus seed valorization and utilization [2]. One study reported the utilization of citrus seeds for fixed oil production, which were enriched with tocopherols, phytosterols, sugars, carotenoids, and minerals [3]. The most commonly detected organic acids in citrus by-products are citric acid, malic acid, oxalic acid, ascorbic acid, fumaric acid, benzoic acid, tartaric acid, and succinic acid. These organic acids act as natural preservatives and powerful antioxidants, protect against oxidative stress, support skin health, and reduce muscle fatigue, anti-inflammatory properties (Table 1).

2 Materials

2.1 Ultrasonic Fluid Extraction

1. Distilled Water: Prepare a solvent by dissolving malic acid (1.34 g for 0.1 M) and fumaric acid (1.16 g for 0.1 M) in distilled water adjust the water and adjust the volume to 100 mL.
2. Dilute HCl (0.1 M): Dissolve 1.921 g of citric acid, 0.900 g of oxalic acid, or 1.501 g of tartaric acid in 0.1 M HCl and adjust the volume to 100 mL with the same HCl solution.

Table 1
Organic acids in citrus by-products

Organic acid	Citrus by-product utilized
Citric acid	Extracted from juice and pulp. Found in vesicles of citrus fruits.
Malic acid	Extracted from juice and pulp. Found in membranes and vesicles of citrus fruits.
Oxalic acid	Present in trace amounts in juice and peels of citrus fruits.
Ascorbic acid	Extracted from juice and pulp. Present in albedo and flavedo of the citrus fruits
Fumaric acid	Found in trace amounts in pulp and juice.
Benzoic acid	Forms during fermentation of citrus waste, especially peels and pulp residues.
Tartaric acid	Found in juice and trace amounts in the pulp of citrus fruits.
Succinic acid	Extracted from fermented citrus peels and pulp residues.

3. Ethanol–Water (1:1): Dissolve 1.761 g of ascorbic acid and 1.221 g of benzoic acid in 100 mL of 1:1 ethanol–water solution for a 0.1 M concentration.
4. Methanol–Water (1:1): Dissolve the required amount of succinic acid (e.g., 0.1 M: 1.18 g in 100 mL) in a 1:1 methanol–water solution.

2.2 Supercritical Fluid Extraction

1. For malic acid; set CO_2 to 31 °C and 1500–3000 psi.
2. For fumaric acid; set CO_2 to 31 °C and 2000–3000 psi.
3. For ascorbic acid; set CO_2 to 31 °C and 1000–2000 psi (low pressure to preserve ascorbic acid).
4. For citric acid; set CO_2 to 35 °C and 2000–3500 psi.
5. For oxalic acid; set CO_2 to 40 °C and 2500–3500 psi.
6. For benzoic acid; set CO_2 to 35 °C and 1000–2500 psi.
7. For tartaric acid; set CO_2 to 35 °C and 1500–3000 psi.
8. For succinic acid; set CO_2 to 35 °C and 2000–3500 psi.

2.3 Solvent Extraction

1. Use 700 mL of 100% ethanol and add 300 mL of water to make 1 L of 70% ethanol; used for extracting polar compounds like ascorbic acid.

   Use 950 mL of 100% ethanol and add 50 mL of water to make 1 L of 95% ethanol; for more concentrated extraction and higher purity acids.
2. Use 700 mL of 100% methanol and add 300 mL of water to make 1 L of 70% methanol; for extracting organic acids like benzoic and tartaric acids.
3. Acetone is often used in 100% concentration as pure acetone; usually used without dilution.

4. pH adjustment agents acidic/alkaline solutions (HCl/NaOH).
5. Drying agents (anhydrous sodium sulfate/magnesium sulfate).

2.4 Enzymatic Extraction

1. Citric Acid: To prepare a 0.5–1% w/v concentration of pectinase for breaking down pectin in citrus peels, dissolve 0.5–1 g of pectinase enzyme in 100 mL of water and stir until fully dissolved.
2. Oxalic Acid: To prepare a 0.5% w/v concentration of cellulase for hydrolyzing cellulose to release oxalic acid, dissolve 0.5 g of cellulase enzyme in 100 mL of water and stir until fully dissolved.
3. Ascorbic Acid: To prepare a 0.1–0.5% w/v concentration of protease for extracting protein-bound ascorbic acid, dissolve 0.1–0.5 g of protease enzyme in 100 mL of water and stir until fully dissolved.
4. Malic Acid: To prepare a 0.5% w/v concentration of amylase for breaking down starch to liberate malic acid, dissolve 0.5 g of amylase enzyme in 100 mL of water and stir until fully dissolved.
5. Fumaric Acid: To prepare a combination of pectinase and cellulase at a 0.5% w/v concentration each for optimal extraction, dissolve 0.5 g of pectinase and 0.5 g of cellulase in 100 mL of water, then stir until both enzymes are fully dissolved.
6. Benzoic Acid: To prepare a 0.1% w/v concentration of protease and 0.5% w/v concentration of pectinase for enhanced extraction, dissolve:
   - 0.1 g of protease in 100 mL of water
   - 0.5 g of pectinase in 100 mL of water

   Then, stir each solution separately until the enzymes are fully dissolved.
7. Tartaric Acid: To prepare a 0.5% w/v concentration of pectinase for the effective release of tartaric acid, dissolve 0.5 g of Pectinase in 100 mL of water and stir until fully dissolved.
8. Succinic Acid: To prepare a combination of cellulase (0.5% w/v) and amylase (0.5% w/v) to target succinic acid precursors, dissolve:
   - 0.5 g of cellulase in 100 mL of water
   - 0.5 g of amylase in 100 mL of water

   Then, combine both solutions and stir thoroughly until both enzymes are fully dissolved.
9. Chemicals: Buffer solution (pH 4–7), acetate, phosphate, or citrate buffers are used.

10. Solvents: Ethanol, methanol, or acetone.
11. Distilled water.

2.5 HPLC

1. Filter the extract to remove solids.
2. Dilute the extract with water or mobile phase if necessary.
3. Use a mixture of water and 0.1% phosphoric acid (dissolve 0.1 g of phosphoric acid in 100 mL of water and stir until fully dissolved) or methanol for separation, with a flow rate; of 0.5–1.0 mL/min.
4. C18 reverse-phase column (4.6 mm × 250 mm, 5 μm particle size).
5. UV detector at 210–220 nm (for most acids). For benzoic acid, detect at 254 nm.
6. Prepare standards of each acid (malic, fumaric, ascorbic, citric, oxalic, benzoic, tartaric, and succinic acid) at known concentrations. Plot a calibration curve for each acid.
7. Inject 10–20 μL of the sample extracted.
8. Measure the peak area or height of each acid and compare with the calibration curve to determine the concentration of each acid in the sample.
9. Each acid has a specific retention time, used for identification and quantification.

2.6 Gas Chromatography

1. To prepare a 0.1–1% w/v derivatize extract, dissolve 0.1–1 g of the extract in 100 mL of solvent.
2. Use helium at a 1–2 mL/min flow rate.
3. Use a nonpolar capillary column.
4. Use an FID (flame ionization detector) detector at 250–300 °C.
5. Prepare standards (0.01–1 mg/mL) and plot a calibration curve.
6. Inject 1–2 μL of sample or standard.
7. Measure peak area and compare with calibration.
8. Use specific retention times for each derivatized acid.

2.7 Spectrophotometer

1. Filter the enzymatic extract to remove solids.
2. If necessary, dilute the extract with distilled water or a suitable solvent.
3. Use specific reagents for each acid:
   - Malic Acid: Use Dihydroxylacetone reagent (specific for malic acid).
   - Fumaric Acid: Use ammonium molybdate.

- Ascorbic Acid: Use 2,6-dichlorophenolindophenol.
- Citric Acid: Use Arsenazo III.
- Oxalic Acid: Use potassium permanganate.
- Benzoic Acid: Use sodium hydroxide for the reaction.
- Tartaric Acid: Use bromothymol blue.
- Succinic Acid: Use sodium tungstate.

4. Wavelength selection:
   - Malic Acid: 230 nm.
   - Fumaric Acid: 350 nm.
   - Ascorbic Acid: 520 nm.
   - Citric Acid: 500 nm.
   - Oxalic Acid: 480 nm.
   - Benzoic Acid: 280 nm.
   - Tartaric Acid: 420 nm.
   - Succinic Acid: 250 nm.
5. Prepare standard solutions for each acid at 0.01–1 mg/mL concentration.
6. Plot a calibration curve for each acid using the absorbance values at their respective wavelengths.
7. Set the spectrophotometer to the appropriate wavelength for each acid.
8. Measure the absorbance of the sample extract.
9. Use the calibration curve to determine the concentration of each acid in the sample based on the absorbance.
10. Compare the sample's absorbance to the standard curve for accurate quantification.
11. This method is simple, cost-effective, and widely used for the quantification of various organic acids in enzymatic extracts.

3 Methods

3.1 Ultrasonic Fluid Extraction

Preparation of Sample

1. Obtain fresh citrus processing residues (peels, seeds, or pulp).
2. Dry the sample in a hot air oven at 50 °C until the moisture is reduced to less than 10%.
3. Pulverize the dried material using a grinder or mortal pestle to achieve a uniform particle size of approximately 0.5 mm.
4. Pass the ground material through a sieve to ensure the uniformity of the sample.

5. Transfer the prepared material into airtight, moisture-free containers for further use.

Procedure for Extraction

1. *Preparation of Sample*
   - Weigh the sample using an analytical balance, and weigh 5 g of prepared citrus powder.
   - Place the sample in a clean beaker.
   - Add 50 mL of prepared solvent. Stir the mixture gently for 2–3 min to ensure the solvent fully saturates the sample.
2. *Ultrasonic Treatment*
   - Set the ultrasonic device, ensuring it is clean and operational.
   - Set Parameters: Frequency (20–40 kHz), power (200–400 W), duration (10–30 min), and temperature (30–50 °C).
   - If using an ultrasonic bath, place the sample container in the bath and make sure it is completely immersed.
   - For ultrasonic probes, submerge the probe in the solvent-sample mixture at 4000 RPM for 10 min [4].
   - Stir intermittently to prevent uneven distribution of ultrasonic waves.
   - Now, filter using vacuum filtration or centrifuge at 5000 RPM for 10 min for finer separation of liquid extract from solid residues.
   - Collect the clear supernatant for further analysis or concentration (optional).
   - For storage, refrigerate at 4 °C for short-term storage or freeze for longer durations.
3. *Postextraction Analysis/Quantification*
   - High-performance liquid chromatography (HPLC): For precise identification and quantification [5].
   - Spectroscopy: Quantify light-absorbing compounds like ascorbic acid at wavelength 265 nm for ascorbic acid.

3.2 Supercritical-Fluid Extraction

Preparation of Sample

1. Fresh citrus fruits or by-products like peels, juice residues, or pulp are chosen as a biomass source.
2. The sample is washed with water to remove dirt, pesticides, and contaminants.

3. Reduce the moisture content to improve the efficiency of SFE, as high moisture can hinder supercritical CO_2 extraction.
4. The dried sample is converted into the powdery form to increase surface area and increase contact with the supercritical fluid. The particle size should be between 0.2 and 2 mm [6].
5. The sample is packed into the extraction chamber to ensure even distribution to maximize the interaction with supercritical CO_2.

Procedure for Extraction

1. Set the extraction vessel to the desired temperature (40–60 °C, depending on the stability of the target acids).
2. Set the pressures of the system to 10:30 MPa (higher pressures increase the solubility of polar acids).
3. Mix a small percentage (5–15% w/v) of ethanol or methanol with the CO_2 to act as a cosolvent, increasing the solubility of polar organic acids.
4. Place the prepared citrus sample into the extraction vessel.
5. Start the flow of CO_2 through the system.
6. Allow the system to run for 30 min to 2 h, depending on the yield and efficiency required.
7. Decrease the pressure in the separator to allow the extracted organic acids to precipitate and collect in a collection vessel.
8. Remove the cosolvent, if used, via evaporation or rotary evaporation.
9. Higher temperatures and pressures may improve extraction but could degrade heat-sensitive acids like ascorbic.

Analysis of Extracted Organic Acids

HPLC: For precise identification and quantification of individual acids (e.g., citric acid, malic acid, and ascorbic acid).

Titration: For total acid content.

UV–Vis Spectroscopy: For specific compounds, like ascorbic acid.

3.3 Solvent Extraction

Sample Preparation

1. Select some citrus by-products like peels, pulp, and juice residues. Wash them with distilled water to remove dirt, debris, and pesticide residues and allow them to air dry.
2. Air drying at room temperature for 24–48 h/oven drying at 40–50 °C to avoid degradation of organic acids [7].

3. Defat if the sample contains fat in excess by using a nonpolar solvent like hexane.
4. Store the sample if the extraction is not immediate; store it in an airtight container in a cool and dry place.

Procedure for Extraction

1. Wash the fruit thoroughly.
2. Peel (if necessary) and homogenize the fruit pulp using a blender or mortar and pestle.
3. Weigh a specific amount of the homogenized fruit pulp (e.g., 10–20 g).
4. Add an appropriate volume of extraction solvent (e.g., ethanol–water, 70:30 v/v) in a 1:5 (sample:solvent) ratio.
5. If required, adjust the solvent's pH to acidic conditions (pH 2–3) using dilute HCl.
6. Stir the mixture vigorously to ensure that the solvent penetrates the sample.
7. Allow the mixture to stand for 30 min to 1 h at room temperature or under mild heating (e.g., 40–50 °C) to enhance extraction.
8. Centrifuge the mixture at 3000–5000 rpm for 10–15 min or filter paper to separate the liquid extract from the solid residue.
9. If needed, use a rotatory evaporator to remove the solvent and concentrate the organic acids.
10. Further purify the extract using liquid–liquid extraction with an organic solvent (e.g., ethyl acetate) or ion-exchange chromatography, depending on the target acid.
11. Quantify the organic acids using analytical methods such as:
    - *Titration*: For total acidity.
    - *HPLC*: For precise identification and quantification.
    - *Spectrophotometry*: For specific acids like ascorbic acid.

3.4 Enzymatic Extraction

1. *Preparation of Sample*
    - Obtain fresh citrus processing residues (peels, seeds, or pulp).
    - Dry the sample in a hot air oven at 40–50 °C until the moisture content is low.
    - Pulverize the dried material into fine powder.
2. *Enzymatic Hydrolysis*

    Buffer preparation:

- *Citric Acid*: Prepare 50 mM phosphate buffer, pH 4.5.
- *Oxalic Acid*: Use 50 mM citrate buffer, pH 4.0.
- *Ascorbic Acid*: Use 50 mM acetate buffer, pH 5.5.
- *Malic Acid*: Prepare 50 mM phosphate buffer, pH 5.0.
- *Fumaric Acid*: Use 50 mM citrate buffer, pH 4.5.
- *Benzoic Acid*: Prepare 50 mM acetate buffer, pH 5.5.
- *Tartaric Acid*: Use 50 mM phosphate buffer, pH 4.0.
- *Succinic Acid*: Prepare 50 mM citrate buffer, pH 5.0.

Dissolve the buffer salts in distilled water adjust pH using HCl or NaOH, and bring to the final volume.

3. *Enzyme Selection and Preparation*
   - Select the enzyme as per the acid and prepare the enzyme solutions in distilled water.
4. *Hydrolysis Setup*
   - Mix the citrus by-product powder with buffer (1:5 or 1:10 w/v ratio).
   - Add the enzyme solution to the substrate–buffer mixture.
   - Incubate at optimal temperature (usually 40–50 °C) for 6–24 h. Stir intermittently.
5. *Extraction of Organic Acid*
   - Centrifuge the hydrolyzed mixture at 10,000 RPM for 10–15 min.
   - Collect the supernatant containing organic acids.
   - Or filter the supernatant through Whatman filter paper or a similar medium.
6. *Postextraction Analysis/Quantification*
   - Perform HPLC [8] OR GC [9] Analysis using appropriate columns and mobile phases for the detection and quantification of acids.

Calculations

Yield of Organic Acids

$$\text{Yield }(\%) = \left[\frac{\text{Weight of extracted acids (g)}}{\text{Weight of dry citrus by-products (g)}}\right] \times 100$$

Concentration of Acids in the Extract

This involves titration *or* spectrophotometric analysis. For titration:

$$\text{Acid conc.}(\text{g/L}) = \left[\frac{\text{Vol.of titrant (L)} \times \text{Normality of titrant (N)}}{\text{Sample vol.(L)}}\right] \times \text{Molecular wt.of acid}$$

Normality

$$\text{Normality (N)} = \frac{\text{Mass of solute (g)}}{\text{Equivalent wt.}\left(\frac{\text{g}}{\text{equiv}}\right) \times \text{Volume of solution (L)}}$$

$$\text{Equivalent weight} = \frac{\text{Molecular weight}}{\text{Number of equivalents}}$$

3.5 Precautions

3.5.1 General Considerations

1. Citrus Peel Handling: Citrus peels contain essential oils that may be co-extracted during the process. Ensure that the extraction conditions do not lead to the unwanted extraction of these volatile oils unless they are desired as part of the product.
2. Citrus Pectin: Pectin, a polysaccharide in citrus, can interfere with the extraction of organic acids, especially during solvent or enzymatic extraction. Adjust the extraction conditions or use specific enzymes (e.g., pectinase) to break down pectin.
3. Storage of Extracts: After extracting organic acids, store the extracts in suitable containers that protect the acids from light, heat, and oxygen. Use amber-colored glass bottles to protect sensitive compounds (like ascorbic acid) from degradation due to UV light.
4. Waste Management: Properly dispose of the leftover biomass, solvents, and enzyme residues according to environmental safety regulations. Citrus by-products should be composted or processed to minimize waste [10].

3.6 For Ultrasonic Fluid Extraction

1. Heat Management: Ultrasound-assisted extraction generates heat due to cavitation. Ensure that the temperature of the extraction solution does not rise excessively, as it may degrade heat-sensitive compounds like ascorbic acid. Use a water bath or cooling system to control temperature [11].
2. Frequency and Intensity Control: The ultrasound frequency and intensity must be optimized. Too high power may destroy sensitive compounds, while insufficient power may result in poor extraction efficiency.
3. Protective Equipment: Wear ear protection during ultrasonic treatment, as high-frequency sound waves can be harmful to hearing if directly exposed to the source.

4. Container Selection: Use containers made of materials that can withstand the mechanical stress of ultrasound waves (e.g., glass or certain plastics). Ensure that the material does not react with the extract and is durable.
5. Avoid Foaming: Citrus by-products can foam during ultrasound extraction, especially due to the presence of essential oils and water. This may reduce the efficiency of the extraction and cause sample loss. Consider using anti-foaming agents.

3.7 For Supercritical-Fluid Extraction

1. High-Pressure Safety: Use of CO_2 as the solvent, which operates under high pressures (above 73.8 atm) and temperatures. Always ensure that the extraction chamber, valves, and connections are well-maintained to avoid leaks and accidents.
2. Avoid Overheating: Citrus by-products may degrade under high temperatures. Control the temperature and pressure to avoid the thermal degradation of sensitive organic acids (like ascorbic acid or citric acid).
3. Material Compatibility: Citrus by-products contain volatile essential oils. Ensure that the extraction system is sealed tightly to avoid the loss of volatile compounds and essential oils [12].
4. Ventilation: Work in a well-ventilated space. CO_2 can displace oxygen, leading to a risk of asphyxiation if proper ventilation is not maintained.
5. Equipment Calibration: Regularly calibrate the supercritical fluid extractor to ensure that pressures, temperatures, and flow rates are within safe and optimal ranges.

3.8 For Solvent Extraction

1. Use of Safe Solvents: While solvents like ethanol or methanol are commonly used for citrus extractions, ensure that the solvents are nontoxic and safe to handle [13].
2. Flammability: Many organic solvents are flammable. Work in a well-ventilated fume hood and keep extraction solvents away from open flames or heat sources.
3. Contamination Prevention: Citrus by-products contain sugars, pectin, and other compounds that may interfere with the extraction of organic acids. Use clean, dry, and contaminant-free glassware and equipment to avoid unwanted reactions or contamination.

3.9 For Enzymatic Extraction

1. Enzyme Selection: Use specific enzymes that can break down the plant cell walls in citrus by-products, releasing organic acids.
2. Temperature Control: Enzymatic reactions are temperature-sensitive. Ensure that the extraction temperature remains within the optimal range for enzyme activity (typically between

30 and 60 °C). Too high temperature may denature the enzyme, while too low may result in ineffective extraction.

3. pH Monitoring: The pH of the extraction medium should be carefully controlled, as enzymes may have a narrow pH range in which they are active. Adjust the pH of the solution to match the enzyme's optimal working conditions.
4. Enzyme Concentration: Use the appropriate concentration of enzymes. Too much enzyme may lead to excessive breakdown of other components, while too little may not yield an efficient extraction [14].

4 Notes

1. Supercritical fluid extraction (SFE) provides the highest yields with minimal solvent usage; on the other hand, UAE and enzymatic extraction are also efficient but may require more time or enzymes.
2. SFE is environmentally friendly due to its use of CO_2, while solvent extraction may involve harmful solvents. Enzymatic extraction is the most sustainable and costly.
3. SFE and UAE offer higher purity and selectivity for organic acids compared to solvent extraction, which may co-extract other compounds.
4. The choice of method depends on the desired application (e.g., food, pharmaceutical, or industrial use), scalability, and specific organic acid targeted for extraction.
5. These organic acids like citric acid, malic acid, and ascorbic acid (vitamin C) are present in abundance in citrus fruits. They are essential for human health by acting as antioxidants.
6. Citrus waste (e.g., peels, pulp, etc.) is a very rich source of organic acids and needs to be utilized [15].
7. The extraction of organic acids from citrus fruits is vital for food, pharmaceutical and green industries, offering sustainable solutions and high-value production [10].
8. Integrating the new techniques, like hybrid methods, enzyme optimization, etc., into zero-waste biomass conversion plants and exploring innovative applications ensures economic as well as environmental benefits.

References

1. Šafranko S et al (2023) Citrus by-products as a valuable source of biologically active compounds with promising pharmaceutical, biological and biomedical potential. Pharmaceuticals 16:1081. Available: https://www.mdpi.com/1424-8247/16/8/1081. Accessed 1 Jan 2025

2. Zayed A, Badawy MT, Farag MA (2021) Valorization and extraction optimization of *Citrus* seeds for food and functional food applications. Food Chem 355:129609. https://doi.org/10.1016/j.foodchem.2021.129609
3. Ahmed RHA, Mariod AA (2022) Citrus seeds fixed oil, composition and its biological activities. In: Multiple biological activities of unconventional seed oils. Academic Press, pp 229–235. https://doi.org/10.1016/B978-0-12-824135-6.00002-7
4. Fernandes FA et al (2022) Recovery of citric acid from citrus peels: ultrasound-assisted extraction optimized by response surface methodology. Chemosensors 10(7):7. https://doi.org/10.3390/chemosensors10070257
5. Nour V, Trandafir I, Ionica ME (2010) HPLC organic acid analysis in different citrus juices under reversed phase conditions. Not Bot Horti Agrobot Cluj Napoca 38(1):44–48. https://doi.org/10.15835/nbha3814569
6. Senevirathne M, Jeon Y-J, Ha J-H, Kim S-H (2009) Effective drying of citrus by-product by high speed drying: a novel drying technique and their antioxidant activity. J Food Eng 92(2):157–163. https://doi.org/10.1016/j.jfoodeng.2008.10.033
7. Li BB, Smith B, Hossain MM (2006) Extraction of phenolics from citrus peels: I. Solvent extraction method. Sep Purif Technol 48(2): 182–188. https://doi.org/10.1016/j.seppur.2005.07.005
8. Lee HS (1993) HPLC method for separation and determination of nonvolatile organic acids in orange juice. J Agric Food Chem 41:1991–1993. Available: https://pubs.acs.org/doi/10.1021/jf00035a033. Accessed 1 Jan 2025
9. Sasson A, Erner Y, Monselise SP (1976) Gas liquid chromatography of organic acids in citrus tissues. J Agric Food Chem 24(3): 652–654. Available: https://pubs.acs.org/doi/10.1021/jf60205a047. Accessed 1 Jan 2025
10. Sharma K, Mahato N, Cho MH, Lee YR (2017) Converting citrus wastes into value-added products: economic and environmently friendly approaches. Nutrition 34:29–46. https://doi.org/10.1016/j.nut.2016.09.006
11. Chemat F, Rombaut N, Sicaire A-G, Meullemiestre A, Fabiano-Tixier A-S, Abert-Vian M (2017) Ultrasound assisted extraction of food and natural products. Mechanisms, techniques, combinations, protocols and applications. A review. Ultrason Sonochem 34:540–560. https://doi.org/10.1016/j.ultsonch.2016.06.035
12. Ray A, Dubey KK, Marathe SJ, Singhal R (2023) Supercritical fluid extraction of bioactives from fruit waste and its therapeutic potential. Food Biosci 52:102418. https://doi.org/10.1016/j.fbio.2023.102418
13. Schlosser Š, Kertész R, Marták J (2005) Recovery and separation of organic acids by membrane-based solvent extraction and pertraction: an overview with a case study on recovery of MPCA. Sep Purif Technol 41(3): 237–266. https://doi.org/10.1016/j.seppur.2004.07.019
14. Marić M, Grassino AN, Zhu Z, Barba FJ, Brnčić M, Rimac Brnčić S (2018) An overview of the traditional and innovative approaches for pectin extraction from plant food wastes and by-products: ultrasound-, microwaves-, and enzyme-assisted extraction. Trends Food Sci Technol 76:28–37. https://doi.org/10.1016/j.tifs.2018.03.022
15. Soares Mateus AR, Barros S, Pena A, Sanches-Silva A (2023) Chapter Two – The potential of citrus by-products in the development of functional food and active packaging. In: Capanoglu E, Navarro-Hortal MD, Forbes-Hernández TY, Battino M (eds) Valorization of wastes/by-products in the design of functional foods/supplements, Advances in food and nutrition research, vol 107. Academic Press, pp 41–90. https://doi.org/10.1016/bs.afnr.2023.06.001

Chapter 10

Extraction, Separation, and Quantification of Naringin in Citrus By-Products Using High Performance Liquid Chromatography (HPLC)

Ankur Kumar and Simran

Abstract

Naringin extraction in the citrus by-products was reported using ultrasonic-assisted extraction (UAE) with a shorter extraction time. Naringin was extracted using a 1/2″ microtip probe at 37% amplitude and a 2 s pulse for 15 min and was eluted at 8.5 min using a 20 min acidic buffer linear gradient. High-performance liquid chromatography (HPLC) with ultraviolet visible detector is the preferred choice for separating polyphenolic compounds due to its selectivity, sensitivity and accuracy over the other colorimetric methods. Major bioactives have a significant absorptivity in the ultraviolet–visible region, making them suitable for HPLC-UV-based separation. In this protocol, we describe a simple, rapid analytical protocol for the reverse phase separation and quantification of naringin using HPLC–UV assisted by ultrasonic extraction with shorter extraction and detection time.

Key words Citrus fruit by-products, High performance liquid chromatography (HPLC), Naringin, Ultrasonic assisted extraction (UAE)

1 Introduction

Citrus fruit juice processing yields by-products, including peel, seed, and pulp. These by-products are reported to have a significant portion of value-added bioactive compounds. Naringin is one of the primary bioactive substances present in citrus species, including oranges, grapes, and lemons. This flavone is also an indicator of bitterness in citrus fruits [1]. Waste peel and seed have been reported to have comparatively higher naringin content than juice extract. Grapefruit peel has been reported to have maximum naringin content in the range of 30–100 mg/g [2, 3], followed by bitter orange (20–80 mg/g) [4, 5] and lemon (1–3 mg/g) [6, 7]. Naringin is widely used in food and nutraceutical applications due to its therapeutic effects. Among other fractions, peel is reported to have the maximum naringin concentration

Shweta Suri et al. (eds.), *Citrus Fruit Processing By-Products: Technologies and Basic Protocols*, Methods and Protocols in Food Science, https://doi.org/10.1007/978-1-0716-5088-2_10,

[8]. Naringin has multiple hydroxyl groups, increasing its antioxidant potential [9]. It is sparingly soluble in water but freely soluble in organic solvents [10].

Naringin extraction is reported using solvent maceration, microwave-assisted extraction (MAE), and ultrasonic-assisted extraction (UAE). Heat-assisted extraction may lead to naringin degradation as it is sensitive to environment, temperature, and light exposure [8, 9, 11]. Ultrasonic-assisted extraction has been reported to have improved yield [9, 12] owing to its better extraction efficiency due to the ultrasonic amplitude used for cell wall disruption [13], enabling the release of naringin into the extraction solvent [12]. Extraction parameters, including time, temperature, ultrasonication amplitude, and choice of extraction solvents, primarily affect the extraction yield. For separation, liquid chromatography is considered one of the most reliable and fast elution techniques due to its selectivity, sensitivity, and reliability, which allows for effective separation even in a complex sample matrix [1, 8]. Naringin is a polar flavonoid [8], resulting in more electronegativity and enhanced polarity, which makes it suitable for separation using high-performance liquid chromatography (HPLC) and the ultraviolet detector [14]. Reverse phase HPLC separation using buffer and organic solvent and octadecylsilane-based stationary phase has been reported for naringin detection [8, 9, 15]. The present protocol reported a simple, rapid reverse-phase HPLC-based gradient method using ultrasonic-assisted sample extraction for naringin quantification in citrus by-products.

2 Material

Prepare all the volumetric solutions using HPLC-grade milli-Q grade water, which has a sensitivity of 18 Ω-cm at room temperature. Use extra pure reagents, including ammonium formate, formic acid, methanol, and acetonitrile. Solvent and HPLC waste shall be handled according to defined disposal provisions.

2.1 Chemicals and Reagents

1. Acetonitrile and methanol, HPLC grade.
2. Ammonium formate, HPLC grade.
3. Naringin, HPLC grade, 99.9% purity.
4. Nylon syringe filter, 0.45 μm.
5. C18, 150 × 4.6 mm, 3 μm HPLC column.

2.2 Equipments

1. Water purification system.
2. Ultrasonic bath type.
3. Binary gradient HPLC system with UV detector.
4. Refrigerated centrifuge with 10,000 rpm capacity.

5. pH meter.
6. Ultrasonication system with 500 W and 20 kHz frequency and 1/2″ microtip.
7. Vortex mixture.
8. Glass vacuum filtration assembly.

2.3 Mobile Phase, Reagent, and Diluent Preparation

1. Diluent: Add 200 mL HPLC-grade water in a 1 L HPLC bottle. Add 800 mL HPLC-grade methanol to make up to 1000 mL. Shake well and sonicate for 5 min using an ultrasonic water bath. Filter through a 0.22 μm membrane filter using a vacuum filtration assembly.
2. Ammonium formate (10 mM), pH 4.00 (*see* **Note 1**): Weigh 0.63 g ammonium formate in 1000 mL HPLC-grade water. Shake well and sonicate for 10 min. Adjust the pH to 4.00 using a 5% formic acid solution.
3. Mobile Phase A: Mix 900 mL 10 mM ammonium formate (*see* **Note 2**) and 100 mL acetonitrile in a 1 L HPLC bottle. Sonicate for 5 min and filter through 0.22 μm membrane filter using a vacuum filtration assembly.
4. Mobile Phase B: Mix 200 mL 10 mM ammonium formate and 800 mL acetonitrile in a 1 L HPLC bottle. Sonicate for 5 min and filter through a 0.22 μm membrane filter using vacuum filtration assembly.

3 Method

Perform all the procedures using an amber-colored volumetric flask at room temperature. Prepare blank, standard, sample, and spiked samples as in Subheading 3.1.

3.1 Blank, Standard, Sample and Spiked Preparation

1. Blank Sample: Use diluent as blank. Filter the solution using a 0.45 μm nylon syringe filter in an HPLC vial and inject.
2. Naringin Standard Preparation: Weigh 10 mg naringin standard (*see* **Note 3**) in a 100 mL amber color volumetric flask. Add 20 mL diluent. Sonicate for 5 min using an ultrasonic bath and make up the volume to 100 mL to make 100 μg/mL standard concentration. Filter the solution using a 0.45 μm nylon syringe filter in an HPLC vial and inject.
3. Sample Preparation Using Ultrasound Assisted Extraction (UAE):
   - (a) Weigh 1 g of citrus by-product dried powder in a 25 mL amber color beaker.
   - (b) Add 15 mL diluent.

(c) Use a 1/2″ diameter microtip at 37% ultrasonication amplitude (*see* **Note 4**).

(d) Keep 2 s pulse mode constant throughout the extraction process.

(e) Continue the extraction process for 15 min using a probe ultrasonicator.

(f) Once the extraction is over, transfer the extract to a 50 mL volumetric flask. Rinse the beaker with the diluent to completely transfer the solution (*see* **Note 5**). Make up the volume to 50 mL using a diluent.

(g) Shake the solution using a glass stopper. Centrifuge the resulting solution using 50 mL centrifuge tubes at 10,000 rpm for 5 min at 10 °C.

(h) Filter the supernatant using a 0.45 μm nylon syringe filter in the HPLC vial and inject.

4. Spiked Sample: Weigh 1 g of citrus by-product powder in a 25 mL amber color beaker and add 10 mL of the standard solution prepared in Subheading 3.1, **step 2** and extract the naringin as in Subheading 3.1, **step 3** and make up the volume to 50 mL using a diluent.

3.2 HPLC Sample Analysis Sequence

Inject blank, sample, standard, and spiked samples (*see* **Note 6**) as per the following sequence and use the mean area value for the result calculation.

1. Column equilibration for 15 min.
2. Column conditioning using gradient elution for 20 min.
3. Blank: 1 injection.
4. Standard Sample: 6 injections (*see* **Note 7**).
5. Blank: 1 injection.
6. Test Sample 1: 3 injections.
7. Test Sample 2: 3 injections.
8. Test Sample 3: 3 injections.
9. Blank: 1 injection.
10. Spiked Sample at 20 μg/mL: 3 injections.

3.3 Naringin Separation and Quantification by HPLC–UV

Separate and quantify naringin in the sample matrix using a binary gradient HPLC system with an octadecyl siloxane (C18) packed LC column. Use the HPLC chromatographic conditions as in Subheading 3.3.1. Naringin peak was eluted at 8.5 min in the test and standard samples. There was no peak in the diluent blank at the naringin retention time. A comparative diluent blank and naringin standard chromatograms are shown in Fig. 1. Further, method suitability and accuracy shall be assessed using a spiked sample (*see*

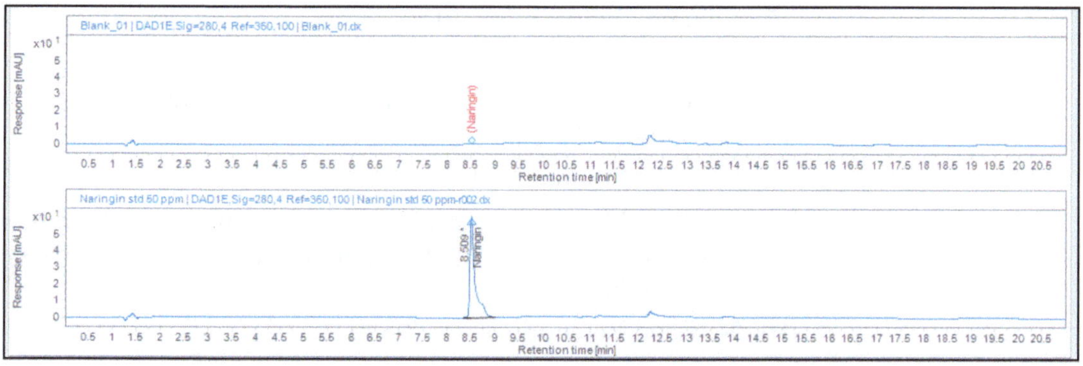

Fig. 1 Comparative diluent blank and naringin standard HPLC chromatogram depicting no blank interference (1) at naringin retention time (2)

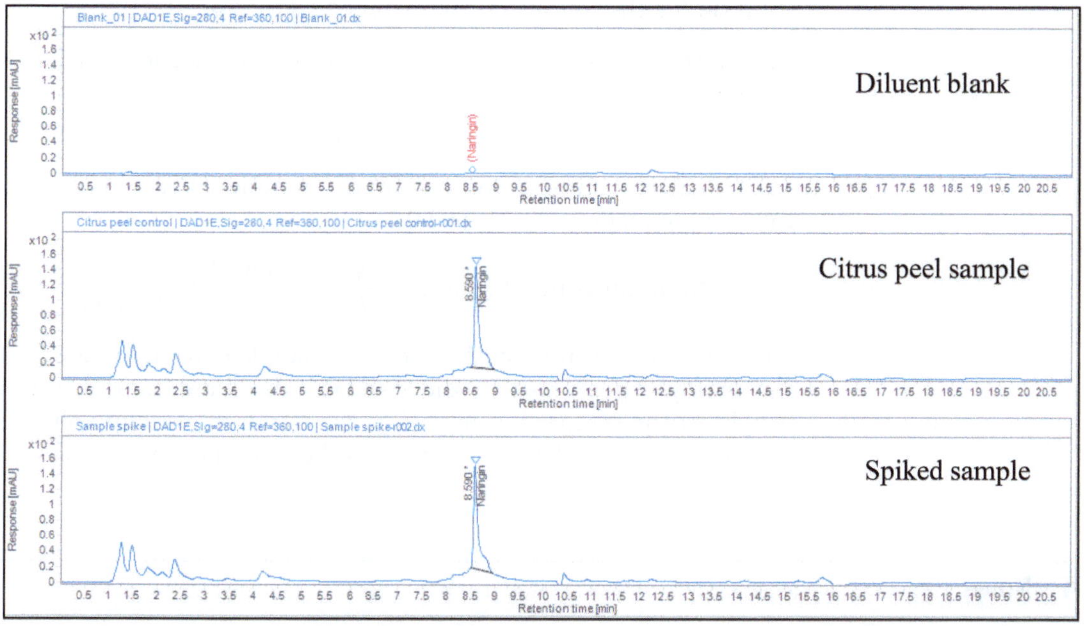

Fig. 2 Representative diluent blank, lemon peel extract, and spiked sample HPLC chromatogram showing the naringin elution pattern in the citrus by-product extract

Subheading 3.1, **step 4**). A representative diluent blank lemon peel extract and spiked sample chromatogram showing the naringin detection are shown in Fig. 2.

3.3.1 Chromatographic Conditions for HPLC Separation

1. HPLC System: Binary gradient (*see* **Note 8**).
2. Flow Rate: 0.8 mL/min.
3. UV Wavelength: 280 nm.
4. Column: C18, 150 × 4.6 mm, 3 μm or equivalent chemistry (*see* **Note 9**).

5. Column Temperature: 40 °C.
6. Sample Temperature: 20 °C.
7. Elution Mode: Gradient elution using mobile phases: A and B (*see* Subheading 2.3).
8. Injection Volume: 5 μL.
9. Gradient Elution: Use gradient used at 0 min (95% A), 5 min (95% A), 7 min (50% A), 10–15 min (5% A), 17 min (95% A), and 20 min (95% A).
10. Naringin Retention Time: 8.5 min.

3.4 Calculations

HPLC chromatographic data assessment is required to check the suitability of the overall analyte peak response used for result calculation. Check the naringin peak system suitability in the standard solution for the following parameters:

1. Relative standard deviation for six replicate injections shall not be more than 1.0%.
2. Symmetry Factor: Shall be less than 1.5.

Calculate naringin content in the sample extract using the following equation (*see* **Note 10**):

$$\text{Naringin content (mg/g)} = \frac{a}{b} \times \frac{c\ \text{(in mg)}}{V_1} \times \frac{V_2}{d\ \text{(in g)}}$$

where a is the naringin peak area in the sample chromatogram; b is the naringin peak area in the standard chromatogram; c is the weight of the standard in mg; V_1 is the standard makeup volume; V_2 is the sample make-up volume; d is the weight of the test sample in g.

4 Notes

1. Adjust buffer pH within ± 0.05 variation from the set value. Calibrate the pH meter before use to measure buffer pH value accurately. A change in pH buffer may result in a significant change in the retention time.
2. Ammonium formate is hygroscopic in nature and shall be kept in a desiccator, and the reagent bottle shall be used at the time of use.
3. Use naringin standard with 99.99% purity and shall be dried before use at around 60–70 °C. If the standard material has less purity, the purity correction factor shall be used to prepare the desired standard concentration.

4. At an amplitude of 37%, the temperature of the extraction solution may increase between 60 and 70 °C. Temperature shall be monitored during extraction using a glass thermometer to avoid overheating.
5. Precautions shall be taken to completely transfer the extraction solution by repeated rinsing with the diluent, as a small droplet loss may cause a significant variation in the sample concentration.
6. A sample with the lowest naringin content, preferably a blank sample matrix without the analyte of interest or the least analyte concentration, shall be used to prepare the spiked sample.
7. Six replicate standard injections help to ensure chromatographic precision. This is measured in terms of the peak area and retention time relative standard deviation.
8. Elution can be performed using a binary or quaternary gradient HPLC system. However, there may be a slight variation in the retention time when using the different HPLC systems and column dimensions. The procedure shall be validated in the laboratory, considering the quality of reagents, solvents, equipment, LC column, and trained personnel.
9. A similar chemistry column can also be used for the single analyte elution. Retention time and system suitability parameters are to be verified for the alternative column.
10. When the working standard was prepared using the standard stock solutions, add all the serial dilutions in the calculation formula.

Acknowledgments

The authors are grateful to the National Institute of Food Technology Entrepreneurship and Management (NIFTEM), Kundli, India, for providing the lab facilities to conduct the experimental work.

References

1. Ribeiro Isabel A, Ribeiro Maria HL (2008) Naringin and naringenin determination and control in grapefruit juice by a validated HPLC method. Food Control 19(4): 432–438, ISSN 0956-7135. https://doi.org/10.1016/j.foodcont.2007.05.007
2. Wilcox LJ, Ashraf-Khorassani M, Taylor LT (2014) Pressurized fluid extraction of flavonoids from grapefruit. J Food Compos Anal 33(1):111–118
3. Tripoli E, Guardia ML, Giammanco S et al (2007) Citrus flavonoids: molecular structure, biological activity, and nutritional properties. Food Chem 104(2):466–479
4. Zhang Z, Peng X, Li S et al (2012) Extraction and identification of flavonoids from Citrus aurantium L. by high-performance liquid chromatography and mass spectrometry. J Agric Food Chem 60(14):3498–3503

5. González-Molina E, Domínguez-Perles R, Moreno DA et al (2010) Natural bioactive compounds of Citrus limon for food and health. J Pharm Biomed Anal 51(2):327–345
6. Xu G, Liu D, Chen J et al (2018) Juice components and antioxidant capacity of citrus varieties cultivated in China. Food Chem 106(2): 545–551
7. Wang Y, Ma Y, Liu J et al (2017) Flavanones and their glycosides in citrus: chemistry and health benefits. Curr Med Chem 24(4): 402–411
8. Sudarshana B, Nilotpal T, Bhupendra S et al (2023) Review on naringin: method of isolation, analytical development, and its recent pharmacological activities. Int J Pharm Sci Res 14(4):1622–1629
9. Csuti A, Sik B, Ajtony Z (2022) Measurement of naringin from citrus fruits by high-performance liquid chromatography – a review. Crit Rev Anal Chem 54:473–486
10. Venkateswara Rao P, Kiran SD, Rohini P, Bhagyasree P (2017) Flavonoid: a review on naringenin. J Pharmacogn Phytochem 6:2778–2783
11. Ioannou I, M'hiri N, Chaaban H et al (2018) Effect of the process, temperature, light and oxygen on naringin extraction and the evolution of its antioxidant activity. Int J Food Sci Technol 53(12):2754–2760
12. Stabrauskiene J, Marksa M, Ivanauskas L et al (2022) Optimization of naringin and naringenin extraction from citrus × paradisi L using hydrolysis and excipients as adsorbent. Pharmaceutics 14(5):890. https://doi.org/10.3390/pharmaceutics14050890
13. Saranya N, Devi P, Nithiyanantham S et al (2014) Cells disruption by ultrasonication. BioNanoSci 4:335–337. https://doi.org/10.1007/s12668-014-0149-2
14. Gorinstein S, Huang D, Leontowicz H et al (2006) Determination of naringin and hesperidin in citrus fruit by high-performance liquid chromatography. The antioxidant potential of citrus fruit. Acta Chromatogr 17:108–124
15. Büyüktuncel E (2017) Fast determination of naringin and hesperidin in natural and commercial citrus juices by HPLC method. Asian J Chem 29:2384–2386

Chapter 11

Analyzing Functional Groups in Citrus By-Products by Fourier Transform Infrared (FTIR) Spectroscopy

Folayemi J. Isaac-Bamgboye

Abstract

This chapter details the protocols for analysis of citrus by-products by using FTIR *via* sample preparation, spectral acquisition, and data interpretation. This includes steps that are essential for enabling the process, such as baseline correction, peak identification, and quantification of functional groups, in a manner that conforms to calibration curves and the Beer–Lambert law. Peak intensity and area calculations are discussed, along with approaches to mitigate the most common challenges (sample interference and atmospheric water vapor); key precautions and best practices, such as sample preparation, instrumentation calibration, and minimizing contamination for reliable results, are discussed.

Key words Fourier transform infrared (FTIR) spectroscopy, Citrus by-products, Functional groups, Spectral analysis, Peak identification, Beer–Lambert Law

1 Introduction

Characterization of functional groups in citrus by-products is essential for their chemical composition, nutritional analysis, and applications in food, pharmaceutical, and cosmetic industries [1].

Bioactive compounds present in citrus by-products (fruit peels, seeds, and pulp residues) are terpenes and limonoids, which play an important role in many biological activities, especially antioxidant, antimicrobial, and anti-inflammatory effects [2]. Rapid nondestructive identification of these by-products is facilitated by advanced spectroscopic methods such as Fourier transform infrared (FTIR) spectroscopy, which can provide a robust platform to valorize these by-products efficiently [3].

Shweta Suri et al. (eds.), *Citrus Fruit Processing By-Products: Technologies and Basic Protocols*, Methods and Protocols in Food Science, https://doi.org/10.1007/978-1-0716-5088-2_11,

1.1 Functional Groups Found in Citrus Fruit By-Products (Terpenes and Limonoids)

A diverse range of bioactive compounds found in citrus by-products are terpenes and limonoids. Hydrocarbons containing characteristic functional groups such as characteristic double bonds and methyl groups are called terpenes, including limonene, α-pinene, and myrcene [4]. There is also a noted contribution of these compounds in the citrus fruits to their aromatics, and they are extensively used in essential oils, food flavoring, and as therapeutic agents [5]. On the other hand, highly oxygenated triterpenoids with lactone structures and hydroxyl groups, namely limonoids, are present in the other compounds [6]. Citrus seeds and peels are rich sources of particular key compounds such as limonin and nomilin. These compounds are shown to have significant bioactivities such as anticarcinogenic and cholesterol-lowering activities [7].

1.2 Principle of FTIR

FTIR makes use of the Fourier transform algorithm that transforms raw interferograms produced in a Michelson interferometer to interpretable spectra [8]. The identification of the various molecular vibrations associated with stretching, bending, and wagging can be achieved based on the great sensitivity of these regions in the FTIR spectrum: fingerprint region (400–1500 cm^{-1}) and functional group region (1500–4000 cm^{-1}). This technique is well-suited to the analysis of complex matrices such as citrus by-products, as it enables the simultaneous detection of multiple functional groups without extensive sample preparation [9]. FTIR instrumentation has recently advanced with a great deal of further simplicity in its use available through the use of attenuated total reflectance (ATR) accessories, allowing for direct measurement of solids and liquids with minimal interference [10].

2 Material

2.1 Citrus By-Products

- *Types of Samples*: The primary samples analyzed are peels, seeds, pulp, and juice residues of citrus by-products.
- *Preparation of Samples*: FTIR analysis requires that the citrus by-products are preprocessed. In general, they are dried and finely ground to a powder of uniform substance for solid samples, which helps with more accurate spectral readings [11]. On the other hand, liquid samples like citrus juices are assayed directly or dried by particular methods to remove excess moisture [12].

2.2 Chemicals and Solvents

- *Solvents for Extraction*: Citrus by-products are extracted using organic solvents such as methanol, ethanol, acetone, and chloroform to remove these specific compounds for FTIR analysis.

- *Reagents for Sample Preparation*: In some instances, reagents including KBr (potassium bromide) are used to bind powdered citrus by-products into pellets for better transmission of infrared light [13].

2.3 FTIR Spectrometer

- *Instrument*: To analyze, a high-quality FTIR spectrometer such as a Perkin Elmer, Thermo Fisher, or Agilent Technologies is required [14].
- *Calibration Materials*: The instrument should be calibrated before any analyses are made using a known standard material such as polystyrene or a similar calibration sample (*see* **Note 1**).

2.4 Glassware and Laboratory Equipment

- *Cuvettes*: To avoid interference in FTIR readings, the cuvettes should be cleaned properly.
- *Mortar and Pestle*: A mortar and pestle is used to grind the solid citrus by-products to a fine powder for grinding.
- *Spatula and Weighing Scales*: Before mixing with solvents or KBr, solid samples should be weighed on a balance or analytical scale [15].

3 Method

3.1 Sample Preparation

The following are the steps for conducting appropriate sample preparation of the different citrus by-products, peels, seeds, juices, and essential oils.

1. *Solid Samples (Peels, Seeds, and Pulp)*:
   - *Cleaning*: Firstly, wash citrus by-products thoroughly to clean dust or dirt, or other pollution that may interfere with the analysis. So, wash the samples using distilled water and ensure that all residues have been removed.
   - *Drying*: For having good spectral results, it is essential to dry as many solid samples as possible before the analysis, especially citrus peels and pulp, to eliminate any moisture content present. Dry the samples until a constant weight is reached, free of excess moisture.
   - *Grinding*: The samples are fully dried and ground in a mortar and pestle, or mechanical grinder, to a fine powder. Fine grinding increases the surface area of the sample and improves the quality of the spectra by eliminating greater interaction with infrared light.
   - *KBr Pellet Preparation*: About 1–2 mg of a ground sample is weighed and then mixed with 100 mg of potassium bromide (KBr). The KBr is transparent to infrared radiation so that it does not interfere with the FTIR measurements. The

mixture is then compressed in a hydraulic press to form a clear and uniform pellet that can be analysed by FTIR.

2. *Liquid Samples (Juices and Extracts)*:
   - *Extraction*: Solvent extraction is performed for liquid samples such as citrus juice or essential oil extracts using suitable solvents, such as methanol, ethanol, and chloroform. If essential oils are to be analyzed, hexane, or diethyl ether should be used to extract the volatile components.
   - *Sample Transfer*: Upon extraction, the liquid extract may then be transferred into a quartz clean, dry cuvette. Always make sure that the cuvette has a proper path length (usually 0.1–1 mm) to get accurate measurements. In case of analysis of essential oil, make sure that there's nothing found floating around to scatter the infrared light.
3. *Intermediate Moisture Samples*:
   - *Homogenization*: Semi-solid or pulp-like citrus by-products can be homogenized in a blender or homogenizer to achieve uniform consistency. The sample may be homogenized, and after drying, if necessary, the FTIR can be used to analyse the sample directly.

3.2 Step-by-Step Procedure for FTIR Analysis

1. *Instrument Setup*: Turn on the FTIR spectrometer, warm it up. For the analysis, set the instrument to the appropriate settings. For example, its spectral range should be from about 4000 up to 400 cm^{-1}, its resolution should be 4 cm^{-1}; and usually 32 or more scans are performed.
2. *Sample Placement*: The solid sample is then placed into the sample holder of the FTIR spectrometer, in place of the KBr pellet for a solid sample. Place the quartz cuvette with the liquid extract in the sample holder for liquid samples. Make sure that the sample is placed correctly so that the light path does not interfere with the sample (*see* **Note 2**).
3. *Background Measurement*: Pinning a background spectrum (usually air or a blank sample) before measuring the sample gives an accounting for any environmental or instrumental interference. When the sample spectrum is subtracted from the background spectrum, a clean result is obtained (*see* **Note 3**).
4. *Spectral Measurement*: Run the FTIR analysis, beginning with IR radiation on the sample and measuring absorbance at each wavelength. An FTIR spectrometer produces a spectrum in which absorbance is reported as a function of wavenumber.
5. *Data Collection*: The FTIR instrument takes multiple scans (usually 32) of the sample to give a better signal-to-noise ratio and to collect high-quality data. The resulting spectrum spans from 4000 cm^{-1} to 400 cm^{-1} (*see* **Note 4**).

6. *Data Analysis*: After collecting the spectrum, perform the data processing steps like baseline correction, smoothing, and normalization to better quality spectrum. Specialized software identifies the key absorption peaks corresponding to specific functional groups in the sample (*see* **Note 4**).

3.3 Calculations

While FTIR analysis primarily provides qualitative data—identifying functional groups in citrus by-products through characteristic absorption peaks—it does not inherently offer quantitative information. However, quantification can be achieved by measuring the intensity of specific absorption bands and correlating them to concentration using calibration curves.

3.3.1 Peak Identification and Functional Group Assignment

The FTIR spectrum displays peaks at specific wavenumbers corresponding to different functional groups. The most common functional groups found in citrus by-products include hydroxyl (–OH), carbonyl (C=O), C–H, and C=C stretches (*see* **Note 3**). Here are some examples of functional group peaks:

- *Hydroxyl (–OH) Stretch*: 3200–3550 cm^{-1}
- *Carbonyl (C=O) Stretch*: ~1725 cm^{-1}
- *C–H Stretch (Aliphatic)*: 2800–3000 cm^{-1}
- *Aromatic C=C Stretch*: 1500–1600 cm^{-1}

3.3.2 Quantitative Analysis

In case of the necessity of quantitative analysis of the sample, the intensity of some peaks can be used to determine the compound concentration. This is done by creating a calibration curve from standard solutions of known concentration. The steps for quantitative analysis are as follows:

1. *Standard Curve Creation*: Standardize the solution of a compound of a given interest at a known concentration. An FTIR absorbance is measured at the specific wavelength related to the functional group of each standard solution.
2. *Regression Analysis*: To construct a calibration curve, plot the absorbance values against the concentrations of the standards.
3. *Sample Analysis*: Conduct the sample absorbance at the same wavelength and use the calibration curve to find the concentration of the compound in the sample (*see* **Note 6**).

3.4 Precautions

To ensure accurate and reliable FTIR analysis of citrus by-products, several precautions must be taken:

1. *Sample Contamination*: Make sure the samples are free of contamination. FTIR spectra can be interfered with by residual solvents, oils, or even moisture. It is necessary to always clean equipment thoroughly between samples in order to avoid contaminating samples.

2. *Moisture Content*: FTIR spectrum interference is possible from excess moisture in a solid sample. To obtain accurate spectral results, samples need to be dried properly (*see* **Note 2**).
3. *Proper Calibration*: Always calibrate the FTIR instrument before each set of measurements to avoid any discrepancies in the spectra. Thereby, the instrument is readable and the publication of the results is reproducible (*see* **Note 3**).
4. *Avoid Sample Overload*: The quantity of the sample used when preparing KBr pellets or liquid samples should be correct. The high sample material can produce saturation of the infrared signal and give distorted spectra.
5. *Correct Sample Positioning*: There is a critical placement of the sample in the FTIR instrument. Solid samples should be placed in the correct spots on the sample holder, and liquid samples need to be put in clean quartz cuvettes. Incorrect or incomplete measurements can occur if the misalignment cannot be corrected (*see* **Note** 7).
6. *Environmental Factors*: Maintain the instrument in a stable environment in which sunlight, temperature fluctuations, and excessive humidity are avoided. It is shown that these factors can affect FTIR results and the reproducibility of measurements (*see* **Note** 7).
7. *Consistent Data Processing*: Make sure that when the analysis is performed, data processing techniques are done consistently, like baseline correction and normalization. It will reduce the variability and increase the reliability of the results (*see* **Note 4**).

4 Notes

1. *Understanding FTIR Spectrum*
   - The FTIR spectrum provides key information about the molecular structure of a sample by displaying the absorption of infrared light at specific wavelengths corresponding to the vibrational modes of functional groups. It is important to recognize the key peaks in the spectrum and understand their relationship to the chemical bonds in the sample.
2. *Interference from Water*
   - Water has a strong absorption band around 3400 cm^{-1}, which can overlap with hydroxyl group absorption bands. If the sample contains moisture, this can distort the FTIR spectrum. Therefore, it is crucial to dry the samples thoroughly to prevent water-related interference in the analysis.

3. *Instrument Maintenance and Calibration*
   - Regular maintenance and calibration of the FTIR spectrometer are crucial to ensure high-quality data. Calibration should be performed using standard reference materials, and the instrument should be checked for alignment, sensitivity, and stability before conducting sample analysis.
4. *Software and Data Interpretation*
   - After acquiring the FTIR spectra, data analysis software is often required to interpret the peaks. Many FTIR instruments come with integrated software that helps identify functional groups by comparing the sample spectra with reference libraries. However, understanding the theoretical basis of FTIR and interpreting spectra manually can provide valuable insights when analysing complex samples.
5. *Functional Group Identification*
   - Citrus by-products, including peels, seeds, and juices, contain various functional groups such as hydroxyl (–OH), carbonyl (C=O), and aromatic rings (C=C). Identifying these functional groups in the FTIR spectrum allows for a better understanding of the chemical composition and potential bioactive properties of the samples.
6. *Significance of Calibration Curve for Quantitative Analysis*
   - When performing quantitative analysis using FTIR, ensure that the calibration curve is constructed using fresh standard solutions to avoid errors due to degradation or contamination. The correlation between absorbance and concentration should be linear within the working range of the instrument for reliable results.
7. *Application to Other Citrus By-Products*
   - The methodology can be applied to various citrus by-products, including not only peels and seeds but also waste products like citrus pulp and residual oils. Each type of sample may require slight modifications in preparation to optimize FTIR results.
8. *Sensitivity to Sample Thickness*
   - For liquid samples, the thickness of the sample in the cuvette affects the absorbance. Ensure that the correct path length is used and avoid overfilling or underfilling the cuvette to ensure accurate measurements [17].
9. *Reproducibility*
   - FTIR is a highly reproducible technique, but factors such as sample handling, instrument calibration, and environmental conditions can influence results. Always ensure consistency in sample preparation and measurement procedures to achieve reliable data [18].

10. *Use of KBr in Solid Sample Analysis*

   - KBr is commonly used to prepare pellets for FTIR analysis of solid samples because it is transparent to infrared radiation. However, KBr absorbs moisture from the air, so it is essential to store it in a dry environment and handle it carefully to avoid interference in the spectra [16].

References

1. Šafranko S, Šubarić D, Jerković I et al (2023) Citrus by-products as a valuable source of biologically active compounds with promising pharmaceutical, biological and biomedical potential. Pharmaceuticals (Basel) 16:1081
2. Wedamulla NE, Fan M, Choi Y-J et al (2022) Citrus peel as a renewable bioresource: transforming waste to food additives. J Funct Foods 95:105163
3. Andrade MA, Barbosa CH, Shah MA et al (2022) Citrus by-products: valuable source of bioactive compounds for food applications. Antioxidants 12:38
4. Saini RK, Ranjit A, Sharma K et al (2022) Bioactive compounds of citrus fruits: a review of composition and health benefits of carotenoids, flavonoids, limonoids, and terpenes. Antioxidants 11:239
5. Shi Y-S, Zhang Y, Li H-T et al (2020) Limonoids from citrus: chemistry, anti-tumor potential, and other bioactivities. J Funct Foods 75:104213
6. Roy A, Saraf S (2006) Limonoids: overview of significant bioactive triterpenes distributed in plants kingdom. Biol Pharm Bull 29:191–201
7. Trache D, Hussin MH, Hui Chuin CT et al (2016) Microcrystalline cellulose: isolation, characterization and bio-composites application—a review. Int J Biol Macromol 93:789–804
8. Ramírez-Hernández A, Aguilar-Flores C, Aparicio-Saguilán A (2019) Fingerprint analysis of FTIR spectra of polymers containing vinyl acetate. Dyna 86:198–205
9. Anderson JM, Voskerician G (2010) The challenge of biocompatibility evaluation of biocomposites. In: Biomedical composites. Woodhead Publishing, pp 325–353. https://doi.org/10.1533/9781845697372.3.325
10. Abdulfatai H, Shoge M, Tamasi A et al (2024) In vitro antimicrobial activity of the oil obtained from the seeds of Citrullus colocynthis (bitter apple). Int J Res Sci Innov 11: 714–732
11. Masyita A, Mustika Sari R, Dwi Astuti A et al (2022) Terpenes and terpenoids as main bioactive compounds of essential oils, their roles in human health and potential application as natural food preservatives. Food Chem 13: 100217
12. Lubinska-Szczygeł M, Kuczyńska-Łażewska A, Rutkowska M et al (2023) Determination of the major by-products of Citrus hystrix peel and their characteristics in the context of utilization in the industry. Molecules 28:2596
13. Mahato N, Sinha M, Sharma K et al (2019) Modern extraction and purification techniques for obtaining high purity food-grade bioactive compounds and value-added co-products from citrus wastes. Foods 8:523
14. Baker MJ, Trevisan J, Bassan P et al (2014) Using Fourier transform IR spectroscopy to analyze biological materials. Nat Protoc 9: 1771–1791
15. Christel S (2016) What cuvette material can be used for measuring many samples of organic (ethyl acetate) solutions? Available at: https://www.researchgate.net/post/What_cuvette_material_can_be_used_for_measuring_many_samples_of_organic_ethyl_acetate_solutions
16. Dai F, Zhuang Q, Huang G, Deng H, Zhang X (2023) Infrared Spectrum Characteristics and Quantification of OH Groups in Coal. American Chemical Society 8(19):17064-17076
17. Huber M, Trubetskov M, Hussain SA, Schweinberger W, Hofer C, Pupeza J (2020) Optimum Sample Thickness for Trace Analyte Detection with Field-Resolved Infrared Spectroscopy. American Analytical Chemistry 92(11)
18. Mirwald J, Khalighi S, Varveri A (2025) Evaluating the reproducibility and consistency of different sample preparation techniques used for ATR-FTIR spectroscopy from the RILEM 295-FBB TG1 round robin test. Mater Struct 58 255. https://doi.org/10.1617/s11527-025-02755-1

Chapter 12

Essential Oil Identification in Citrus Seeds

Sayani Mavai and Rajni Chopra

Abstract

Citrus seeds, a significant by-product of the fruit processing industry, remain an unexploited resource for essential oils with remarkable techno-functional and economic potential. These oils are rich in bioactive compounds such as aldehydes, terpenes, and esters, known for their antioxidant, therapeutic and antimicrobial characteristics, making them valuable in pharmaceutical and food processing sectors. This chapter explores extraction methods, including solvent-based techniques, and highlights advanced analysis tools like gas chromatography (GC), gas chromatography–mass spectrometry (GC–MS), high-performance liquid chromatography (HPLC), and enzyme-linked immunosorbent assay (ELISA). Furthermore, the key precautions ensuring precision and reliability during the extraction and analysis are also discussed. By emphasizing sustainable practices and resource efficiency, this chapter reveals the potential of citrus seed waste as a valuable resource for diverse industrial applications. Hence, this approach not only promotes waste valorization but also contributes to economic and environmental sustainability in the food processing sector.

Key words Citrus seeds, Essential oil, Extraction, GCMS, HPLC, ELISA

1 Introduction

Citrus fruits, part of the Citrus genus in the Rutaceae family, are grown globally for their economic and nutritional benefits [1]. While the pulp and juice of these fruits are commonly utilized, the seeds are frequently discarded. However, recent research indicates that citrus seeds can serve as a valuable source of essential oils, which are complex mixtures of volatile aromatic compounds. These oils are known for their functional properties, including antimicrobial, antioxidant, and therapeutic effects, making them highly sought after in various industries such as pharmaceuticals, cosmetics, and food processing [2].

Extracting essential oils from citrus seeds offers a chance to convert agricultural waste into valuable resources, promoting waste reduction and sustainable practices [3].The extraction and analysis of these oils require advanced techniques to identify their chemical

Shweta Suri et al. (eds.), *Citrus Fruit Processing By-Products: Technologies and Basic Protocols*, Methods and Protocols in Food Science, https://doi.org/10.1007/978-1-0716-5088-2_12,

components, which include terpenes, aldehydes, esters, and other bioactive compounds. These compounds are associated with a variety of biological activities, such as antimicrobial, antifungal, and antioxidant effects. Furthermore, using these essential oils presents a cost-effective method to improve resource efficiency in the citrus-processing industry [4].

Accurate and sophisticated analytical methods are essential for the reliable identification and quantification of essential oils extracted from citrus seeds [5]. Techniques like gas chromatography (GC) and gas chromatography–mass spectrometry (GC–MS) are frequently used to separate and analyze the volatile components, providing detailed chemical profiles of the oils. High-performance liquid chromatography (HPLC) is especially effective for detecting and quantifying nonvolatile constituents, offering insights into the overall composition. Additionally, enzyme-linked immunosorbent assay (ELISA) acts as a biochemical tool for identifying specific bioactive compounds or molecular markers present in the oils. The combination of these analytical methods ensures a thorough evaluation of the chemical properties and potential uses of citrus seed essential oils [6].

In this chapter, we will discuss the various essential oils extracted from citrus fruit seeds, focusing on their applications in the food industry. It will also explore the different analytical techniques employed to estimate and analyze these oils, including GC, GC–MS, HPLC, and ELISA. The chapter will outline the necessary materials and reagents, followed by a detailed methodology that includes sample preparation and a step-by-step procedure. Additionally, it will cover the assessments and calculations needed for analyzing the results, along with essential precautions to ensure accurate outcomes. Through this organized approach, the chapter aims to provide a thorough understanding of the methods and applications of essential citrus seed oils.

1.1 Essential Oils in Citrus Fruit Seeds

Citrus seed-derived essential oils are a promising and valuable resource for the richness of various bioactive compounds, with great potential for applications in different areas [7]. The main composition is volatile aromatic compounds that are a secondary metabolite result, as observed in Fig. 1. Terpenes, such as pinene, linalool, and limonene, are responsible for the characteristic odour and therapeutic properties of the citrus essential oils. Citrus seed oils have gained significant attention based on their uniqueness in chemical composition, as lemon seeds contain the highest oil yield (78.9%), sweet orange (51.8%) and bitter orange and bergamot possess 26% oil content. Among these, linoleic acid is the dominant fatty acid present, especially in the seed oils of mandarin, sweet orange, and grapefruit. These oils also possess a considerable amount of unsaturated fatty acids, such as linoleic acid and oleic acid, distinguished for their antimicrobial, antioxidant, and anti-inflammatory characteristics [8].

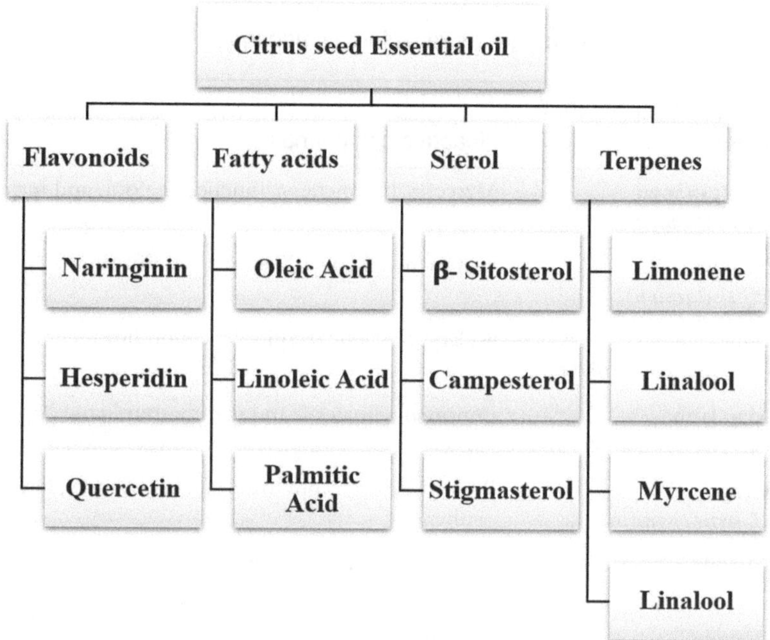

Fig. 1 Chemical constituents of citrus seed essential oil

Citrus seed oils are rich in sterols, specifically β-sitosterol, the most abundant sterol in grapefruit seed oil (Table 1). Sterols are more stable than fatty acids and thus contribute to the overall stability of the oil. The emerging studies have pointed out the antioxidant properties of citrus seed oils, whereby some species, including grapefruit, musami lime, and kinnow mandarin, have antioxidant activities that are better compared to conventional vegetable oils like sunflower, cottonseed, and groundnut [9].

Since over 130 million metric tons of citrus are produced globally annually, citrus seeds, once regarded as waste products, are being used more to obtain essential oil. This gives the agricultural waste an added value while promoting sustainability and environmentally friendly citrus production practices. With continued research into the phytochemical diversity and health benefits of citrus seed oils, their applications both commercially and industrially have continued to grow, driving innovation in natural product industries.

1.2 Applications in Food/Nonfood Sector

Citrus seed essential oils, sourced from fruits such as *Citrus aurantiifolia* (lime), *Citrus limon* (lemon), and *Citrus sinensis* (sweet orange), are rich in bioactive compounds, including terpenes such as aldehydes, pinene, linalool, esters, and limonene [25]. The compounds are contributing to the distinct aromatic properties of the oils, besides their therapeutic potential. Citrus seed oils are used primarily as natural flavoring agents that improve the sensory

Table 1
Sources of citrus fruits and its bioactive compounds

Sources of citrus fruit seeds	Bioactive compounds	References
Sweet orange (*Citrus sinensis*)	Myrcene, limonene, α-pinene, linalool, and terpinene	[9, 10]
Lemon (*Citrus limon*)	Limonene and acyl-N ω-methylserotonins	[11, 12]
Lime (*Citrus aurantiifolia*)	β-Caryophyllene, terpinolene, and limonene	[11, 13]
Mandarin (*Citrus reticulata*)	Limonene, linalool, and gamma-terpinene	[14, 15]
Grapefruit (*Citrus paradisi*)	Proanthocyanin, phytosterol, tocotrienols, flavonoids, and phenolic acids	[16]
Citron (*Citrus medica*)	Limonoids and Limolene	[17, 18]
Pomelo (*Citrus maxima*)	Naringenin, hesperidin, and neohesperidin	[19–21]
Tangerine (*Citrus tangerine*)	d-Limonene and γ-terpinene	[22]
Bergamot (*Citrus bergamia*)	Bergapten, linalool, α-pinene, and β-pinene	[23, 24]
Kaffir lime (*Citrus hystrix*)	Citral, γ-terpinene, limonene, β-pinene, myrcene, and linalool	[25]
Sweet lemon (*Citrus limon*)	Campene, myrcene, citral, β-pinene	[26]

profiles of products like baked goods, beverages, sauces, and confectionery, in the food sector. They are also valued in food preservation because of their antimicrobial properties that help inhibit the growth of *Listeria monocytogenes*, *Aspergillus flavus*, *Salmonella* spp., and *Escherichia coli*. Antimicrobial action is primarily attributed to terpenoid compounds, particularly limonene, which disrupt microbial cell membrane integrity. This disruption causes the release of cellular contents, ultimately resulting in cell death. Additionally, the citrus seed oils exhibit potent antioxidant activity due to the presence of phenolic and flavonoid compounds. These antioxidants neutralize free radicals, thereby preventing the oxidation of lipids and maintaining nutritional value, shelf life of food and freshness of the products [17, 27].

Furthermore, citrus seed essential oils are increasingly gaining attention in the non-food sector, especially in the pharmaceutical and nutraceutical sectors [28]. These oils contain bioactive compounds like linalool and limonene, which possess antimicrobial, anti-inflammatory, and analgesic characteristics. Moreover, they are utilized in the nutraceutical sector as dietary supplements designed to reduce inflammation and oxidative stress, hence playing a significant role in managing chronic conditions such as arthritis and cardiovascular diseases. The bioactive compounds present in essential oil inhibit the pro-inflammatory pathways, including the reduction of cyclooxygenase-2 (COX-2) and inflammatory cytokines [29]. Additionally, their capacity to support wound healing, enhance skin health, and restrain bacterial growth further highlights their therapeutic promise.

1.3 Various Types of Estimation Techniques

1.3.1 Gas Chromatography (GC)

GC is an effective analytical technique used to separate volatile components based on their interaction affinity with a stationary phase and a mobile gas phase [30]. Initially, the sample is introduced into a heated injector, where it transforms into a gas. The resulting vapors are further transported through a chromatographic column by an inert gas, such as helium, hydrogen or nitrogen. Compounds that are more volatile or have weaker interaction affinity with the stationary phase move through the column more quickly, while those with lower volatility or have stronger interactions are held back longer. This deviation in retention time enables effective separation and identification of the individual compounds. GC technique aids in the analysis of the volatile compounds of citrus seed essential oils, which contain a wide range of bioactive and aromatic compounds. The identification of monoterpenes, such as pinene, limonene, sesquiterpenes, and oxygenated components, such as citral and linalool, can be identified and quantified accurately through this technique. Additionally, GC aids in the formulation of a detailed chemical profile by comparing the retention times of the analytes with those of known reference standards, hence facilitating the precise identification of specific components like limonene [31, 32].

1.3.2 Gas Chromatography–Mass Spectrometry (GC–MS)

GC–MS is an analytical methodology that merges the separation capabilities of gas chromatography (GC) with the molecular identification function of mass spectrometry (MS). In the GC phase, volatile and semi-volatile compounds are separated based on their physicochemical properties, such as volatility and polarity, as they interact with the stationary phase of the chromatographic column. Once separated, these compounds are introduced into the mass spectrometer, where they undergo ionization, typically through chemical ionization or electron impact. This ionization process fragments the [33]. The resulting mass spectrum acts as a molecular fingerprint, allowing for the precise identification and quantification of individual compounds by comparing them with established spectral libraries or known standards. GC–MS is preferred for its ability to analyze complex compounds in detail, making it especially effective for profiling the bioactive components in citrus seed essential oils. This method offers both sensitivity and specificity, enabling the identification and quantification of both major and trace constituents with high accuracy. Citrus seed essential oils contain a variety of bioactive compounds, including monoterpenes (like β-myrcene, limonene, and α-pinene), sesquiterpenes, aldehydes (such as citral), esters, and phenolic compounds. GC–MS facilitates a thorough characterization of these constituents, providing a comprehensive understanding of the chemical profile. For example, limonene, a prominent monoterpene, can be quantified alongside lesser components like neral or geranial, which are linked to antimicrobial and antioxidant activities. Such detailed profiling is essential for uncovering the functional properties of the oil and assessing its potential applications [34]. By providing detailed and accurate data, GC–MS aids in the exploration of citrus seed essential oils for various applications in the food industry.

1.3.3 High-Performance Liquid Chromatography (HPLC)

HPLC, an advanced analytical methodology designed to separate, identify, and quantify compounds that are nonvolatile or thermally labile. The method uses a liquid mobile phase to transport the sample through a high-pressure column packed with a stationary phase, where the separation is based on the differential interactions that the analytes have with the two phases, depending on their chemical properties, such as polarity, size, and solubility of the molecules [35]. Analytes that have a weaker affinity for the stationary phase elute faster and those that are strongly interacting require more time to pass through the column. The detection systems include ultraviolet–visible (UV–Vis) spectroscopy or fluorescence detectors, which monitor the eluted compounds and give qualitative and quantitative data. HPLC is one of the analytical tools used in the characterization of the nonvolatile bioactive compounds in citrus seed essential oils. These are mainly phenolic compounds, flavonoids, and other polar constituents that determine the functional properties of oils [36].

The citrus seeds essential oils include significant amounts of flavonoids and phenolic acids in citrus seed essential oils like hesperidin, rutin, and naringin, with potent antioxidants and antimicrobial properties. They are separated using HPLC for accurate determination based on retention time and peak characteristics. This detailed profiling is crucial for understanding the chemical composition of the oil and its potential bioactivity in various applications [37]. HPLC also plays an important role in the quality control and standardization of citrus seed essential oils. The technique ensures uniformity across different batches or sources of oil by providing a consistent chemical fingerprint. Especially about applications in foods, nutraceuticals, and pharmaceutical applications, the standardized content of active ingredients determines efficiency. HPLC also finds another vital application in indicating impurities and degradation products, thereby ensuring safety in quality for commercial and industrial consumptions. Thus, HPLC can be used to provide very accurate and reproducible results for the advancement of knowledge in the utilization of citrus seed essential oils for various applications.

1.3.4 Enzyme-Linked Immunosorbent Assay (ELISA)

ELISA, an ultrahigh analytical technique based on the antigen–antibody interaction principle [38]. This technique depends on the usage of antibodies whose reaction is specific to a target compound (antigen) found within a sample. After the occurrence of the bonding, a conjugated enzyme by a secondary antibody is introduced in the mixture, which catalyses a reaction on a substrate; this leads to the formation of a detectable signal, most typically colorimetric, fluorometric, or chemiluminescent. The signal intensity is directly linked to the concentration of the target compound in the sample, thus allowing for its precise quantification. ELISA is an innovative and effective technique for the analysis of specific bioactive components in citrus seed essential oils, particularly those present in trace amounts. Citrus seed essential oils are known to contain limonoids (e.g., limonin and nomilin), flavonoids (e.g., naringin and hesperidin) and are associated with their antimicrobial, antioxidant, and anti-inflammatory properties. It is therefore crucial to identify and quantify these compounds to assess the quality and bioactivity of essential oils. ELISA is especially useful for guaranteeing product consistency and quality. The process validates the existence and quantity of bioactive compounds, thus supporting standardization in the commercial production of essential oils for use in nutraceuticals and pharmaceuticals. Additionally, ELISA can be applied in research to monitor the effect of processing and extraction methodologies on the retention of key bioactive components.

The specificity of ELISA ensures that only the desired compounds are measured, reducing the potential for interference from other constituents. This level of precision is vital for linking the presence of bioactive compounds to the therapeutic and functional characteristics of citrus seed essential oils. Overall, ELISA serves as a critical tool for advancing the understanding and application of these oils in health-related fields.

2 Materials

2.1 Materials and Reagents Required

Citrus seeds used in the analysis should be obtained from mature citrus fruits. Cleaning is also an essential part of removing residual pulp, peel, or other contaminants; hence, seeds can be cleaned by a thorough running distilled water stream and dried further. It is essential that the seeds be handled cautiously during this process, so that any physical damage during the process will not cause the oil to be prematurely released and ensure that its chemical profile is well preserved for the subsequent extraction and analysis [39, 40].

2.2 Solvents and Reagents

2.2.1 Hexane or Diethyl Ether

1. Analytical grade solvents are mandatory for oil extraction to ensure purity and accuracy.
2. Hexane (preferred due to its low boiling point) or diethyl ether should be stored in sealed containers to prevent evaporation and cross-contamination.
3. Store in a cool, dry place with a temperature range of 15–25 °C.

2.2.2 Sodium Sulfate (Anhydrous)

1. Essential for drying the extracted oil by removing residual moisture.
2. Must be pre-activated by heating to 120 °C for 2 h and stored in desiccators to maintain its anhydrous state.

2.3 Equipment and Glassware

2.3.1 Equipment and Setup

1. Mortar and pestle.
2. Soxhlet extractor.
3. Rotary evaporator.

2.3.2 Glassware

1. Vials.
2. Pipettes.
3. Funnels.

3 Method

The identification and characterization of citrus seed essential oils demand a very defined, structured methodology to result in reliable determinations. Generally, the method can be segregated into four stages of sample preparation, such as oil extraction, removal of solvent, and chemical analysis, as observed in Fig. 2. Therefore, each must be carried out with precision, which would ensure the integrity of the essential oils and ensure valid outcomes in analyses conducted subsequently [41, 42].

3.1 Sample Preparation

3.1.1 Collection and Cleaning of Seeds

This involves the selection of thoroughly matured citrus fruits, from which the seeds are obtained. Primarily, the seeds are washed meticulously to remove all residual pulp, peel, and other materials that have the potential to contaminate the essential oils. It is done by washing the seeds in distilled water, followed by drying, to remove the excess moisture. The citrus seeds require care to ensure that they don't suffer mechanical damage, which in turn may facilitate the release of essential oils without their being allowed to be appropriately extracted [43].

Fig. 2 Systematic extraction and analysis of citrus seed essential oil

3.1.2 Drying and Grinding of Seeds

The seeds are dried once cleaned to maintain the moisture content less than 5%. Drying can be achieved by placing the seeds in a hot air oven at 105 °C for 3 h. Subsequently, the seeds are ground into smaller particles, with the aid of pestle and mortar. Grinding enhances the surface area of the seeds, hence improving the essential oil extraction [44].

3.2 Essential Oil Extraction

3.2.1 Soxhlet Extraction Method

The Soxhlet extraction technique is utilized, as it ensures maximum oil yield through a continuous solvent reflux system. The ground seeds are kept in a thimble, and an appropriate solvent, such as diethyl ether or hexane, is used to extract the essential oils. Hexane is usually preferred due to its low boiling point and efficiency in dissolving essential oils [45].

3.3 Solvent Removal by Rotary Evaporation

Subsequently, the solvent content is removed to separate the concentrated fraction of essential oils of citrus seeds. This procedure is carried out with the aid of a rotary evaporator, allowing the gentle solvent removal under lower pressure. Afterwards, transfer the extract into the rotary flask and adjust the system to a low temperature, ensuring it does not exceed 40 °C to avoid the potential loss of delicate, heat-sensitive volatile compounds during the procedure [46].

3.4 Analysis of Essential Oil

3.4.1 Analysis by GCMS

1. To prepare the essential oil for analysis, first dilute 0.1–1 mL of the essential oil with 1–5 mL of hexane or diethyl ether in a clean container. Ensure the solution is mixed thoroughly to achieve uniformity. Carefully transfer the diluted sample into

amber vials (1.5–2 mL) with airtight caps to prevent exposure to light, which can degrade sensitive volatile compounds [32].

2. Configure the gas chromatograph with a nonpolar capillary column, such as HP-5 or DB-5, measuring 30 m in length, 0.25 mm in internal diameter, and having a film thickness of 0.25 μm. Utilize helium gas of at least 99.99% purity as the carrier, setting the flow rate between 1 and 2 mL/min, depending on the instrument specifications. Program the temperature gradient starting at 50 °C with a hold time of 2 min, gradually increasing at a rate of 10 °C/min, and terminating at 250 °C, where the temperature is held for an additional 5–10 min. For injection, use 1–2 μL of the prepared sample, operating in split mode for concentrated samples or splitless mode for lower concentrations to enhance sensitivity [5, 32].
3. The prepared sample is injected into the gas chromatograph's injector port, where it is vaporized and directed into the column via the helium carrier gas. As the sample travels through the column, its components are separated based on their interaction with the stationary phase and their respective volatilities. Upon exiting the column, the compounds are introduced into the mass spectrometer. The compounds are broken down into positively charged ions using electron ionization at 70 eV. These ions then have their mass-to-charge ratios determined in the mass analyzer, such as in a quadrupole or time-of-flight system, to give a separate mass spectrum for each compound [5].
4. Subsequently, estimate the concentration of citrus seed essential oil by using the formula:

$$\text{Concentration of the analyte} = \left[\frac{A_{\text{sample}}}{A_{\text{standard}}} \times C_{\text{standard}}\right] \quad (1)$$

where A_{standard} = area under the peak for the standard, A_{sample} = area under the peak for sample, and C_{standard} = Concentration of the standard.

3.4.2 Analysis by HPLC

1. Under sample preparation, the step that is essential measures 0.1–1 mL of the crude essential oil by a graduated cylinder or burette; add to this diluted quantity a range of 1–5 mL with an appropriate solvent like acetonitrile, water or methanol, in conjunction with the organic solvent; all this, compatible with the column as well as mobile phase. The sample is filtered through a 0.22 μm or 0.45 μm filter to remove particulate matter and transferred to a vial to prevent light degradation and evaporation. The sample is filtered through a 0.22 μm or 0.45 μm filter to remove particulate matter before transferring it into an amber HPLC-grade vial to prevent light evaporation and degradation [47].

2. Utilize A C18 reversed-phase column, for carrying out compound separation during the setup and analysis of instruments. The mobile phase such as acetonitrile, methanol, or water can be used. It is degassed before filtration. Typical flow rate is in the range of 0.8–1.5 mL/min. UV–Vis detection is normally performed at 280 nm [47, 48].
3. After the chromatographic runs, the peaks are identified and quantified by relating retention times to standards. The integration of peak areas along with a calibration curve is done for quantification. In more complicated cases, molecular weight and structural information may be obtained through HPLC–MS coupling. After the analysis, the column should be flushed with suitable solvents for continued performance [48].
4. Hence, assessment of the concentration is performed by using the following equation:

$$\text{Concentration of the analyte} = \left[\frac{\text{Ab}_{\text{sample}}}{\text{Ab}_{\text{standard}}} \times C_{\text{standard}}\right] \quad (2)$$

where $\text{Ab}_{\text{sample}}$ = Absorbance of the sample, $\text{Ab}_{\text{standard}}$ = Absorbance of the standard, and C_{standard} = Concentration of the standard.

3.4.3 Analysis by ELISA

1. During the sample preparation phase, dilute the essential oil fraction in phosphate-buffered saline (PBS) to ensure the concentration of the target analyte is within the detection range of the specific ELISA kit. The sample must be filtered to remove any particulates that may interfere with the assay [49].
2. Introduced the prepared sample into the wells of the microplate, where the analyte binds to the immobilized antibody. After incubation, the unbound substances are removed through washing steps. Following this, a secondary antibody, conjugated with a horseradish peroxidase or alkaline phosphatase, is added. This secondary antibody binds to the target analyte and catalyzes the conversion of a substrate to a detectable product. The substrate reacts with the enzyme to produce a colour change, and the intensity of this colour is directly proportional to the concentration of the target analyte present in the sample [49].
3. The final data analysis step involves measuring the absorbance of each well using a microplate reader, typically at a wavelength of 650 nm. The concentration of the analyte is determined by comparing the measured absorbance values to a standard curve prepared from known concentrations of the target compound [50].
4. Subsequently, calculate the concentration of the analyte, by utilizing the following formula:

$$\text{Concentration of the analyte} = \left[\frac{(A_{\text{sample}} - A_{\text{blank}})}{(A_{\text{standard}} - A_{\text{blank}})} \times C_{\text{standard}}\right] \quad (3)$$

where A_{sample} = Absorbance of the sample, A_{blank} = Absorbance of the bank, $A_{standard}$ = Absorbance of the standard, and $C_{standard}$ = Concentration of the standard.

3.5 Precautions

1. During extraction, ensure the use of high-purity solvents such as hexane or diethyl ether to avoid contamination. Plant material should be free of surface contaminants, and care should be taken not to damage the plant, which could affect the quality of the essential oil. Extracted oils should be stored in amber vials to protect against light degradation and evaporation [51].
2. When preparing samples for HPLC or GC–MS analysis, ensure proper filtration through a 0.22 μm or 0.45 μm filter to prevent particulate matter from interfering with the analysis. For ELISA, filter samples are used to remove any contaminants that could interfere with the immunoassay [52].
3. For HPLC, use high-quality C18 reversed-phase columns and ensure that the mobile phase is degassed and filtered to avoid air bubbles or contaminants: set flow rates and temperatures according to the specific column and sample properties. Regularly calibrate the detectors and equipment for consistent results [52].
4. In GC–MS, ensure the correct column type (nonpolar, like HP-5 or DB-5) is used and maintain the injector and ion source to avoid contamination. Always use high-purity helium (≥99.999%) as the carrier gas. Calibration of the mass spectrometer is essential for accurate mass-to-charge ratio measurements [53].
5. Accurate results depend on precise sample injection, consistent temperatures, and correct detection wavelengths. For HPLC, carefully monitor retention times and peak areas, while for GC–MS, maintain proper ionization conditions and ensure efficient separation of volatile compounds. In ELISA, proper incubation times and washing steps must be kept, avoiding background interference and providing accurate quantification [53].

4 Notes

1. During the extraction process of essential oils, it is necessary to use fresh plant material, free from any impurities, to ensure the quality of the extracted oil [54]. The selection of the solvent is performed based on the polarity of the essential oil to be extracted and on the requirements of the subsequent analytical technique [55].

2. Generally, a 0.22 μm or 0.45 μm filter is necessary for dilution and filtration of samples, as it removes all particulate matter, which would cause blockage in HPLC or GC–MS columns [56]. Moreover, HPLC-grade reagents should be used to avoid any contamination. The mobile phase in the HPLC analysis must be degassed as air bubbles within it will cause instability in flow and fluctuations in pressure [57].
3. While performing HPLC analysis, the types of columns, such as a C18 column, should be selected. The mobile phase must be designed to achieve optimum separation of the compounds, and the flow rate should be adjusted according to the characteristics of the compounds and the column specifications. The HPLC, GC–MS, and ELISA systems need periodic calibration to ensure that results are accurate and reliable. Moreover, the standards should always be prepared freshly [57, 58].
4. Finally, the obtained essential oils should be stored in amber glass vials with airtight caps immediately after extraction to protect them from temperature, air and light. For long-term storage, refrigeration is recommended to maintain the chemical stability of the citrus seed essential oil [59]. To accurately identify compounds during the analysis of peaks obtained using HPLC or GC–MS, retention times and mass spectra should be compared with those of established standards. Additionally, proper handling is crucial during the performance of ELISA to avoid contamination, and incorporating suitable control samples is crucial to ensure accurate results [50, 60].

References

1. Palangasinghe PC, Liyanage WK, Wickramasinghe MP, Palangasinghe HR, Shih H-C, Shiao M-S, Chiang Y-C (2024) Reviews on Asian citrus species: exploring traditional uses, biochemistry, conservation, and disease resistance. Ecol Genet Genom 32:100269. https://doi.org/10.1016/j.egg.2024.100269
2. Agu CM, Ani KA, Ani ON, Chinedu MP, Kadurumba CH, Ahaneku IE (2024) Waste orange seeds (Citrus sinensis seed) transformation, a viable industrial bio-oil: oil extraction, physicochemical characterization and vital instrumental analyses studies. Waste Manag Bull 1:134–142. https://doi.org/10.1016/j.wmb.2023.10.008
3. Kumar P, Harikrishnan R, Jeevakanth V, Kapoor A, Pal DB, Raghunathan M (2025) Extraction and purification of essential oils from food waste-derived waste biomass. In: Fungal waste biomass management for energy, environment and value-added products. Springer, Cham, pp 301–321
4. Dhotre I (2025) A comprehensive review on progression and innovations in microwave assisted extraction technology for essential oils. J Chem Technol Biotechnol 100(5): 894–907
5. Gogoi R, Sankar Das S, Deka S, Talukder A, Sahariah BJ, Dua TK, Dutta KN (2025) Determination of chemical composition, antimicrobial activity and in silico investigation of the essential oil extracted from peels of Citrus medica (Jora Tenga) & Citrus jambhiri (Gol Nemu) grown in north eastern region of India. Pharmacol Res Nat Prod 6:100139. https://doi.org/10.1016/j.prenap.2024.100139
6. Hegazy MM, Mekky RH, Ibrahim AE, Abouelela ME, Kedra TA, Al-Harrasi A (2025) Essential oils: the science of extraction and its implications for composition and biological activity—a review. Food Anal Methods 18: 1483–1513. https://doi.org/10.1007/s12161-025-02808-9

7. Fernández-Cabal J, Avilés-Betanzos KA, Cauich-Rodríguez JV, Ramírez-Sucre MO, Rodríguez-Buenfil IM (2025) Recent developments in Citrus aurantium L.: an overview of bioactive compounds, extraction techniques, and technological applications. Processes 13: 120
8. Mateus ARS, Teixeira JD, Barros SC, Almeida C, Silva S, Sanches-Silva A (2024) Unlocking the potential of Citrus medica L.: antioxidant capacity and phenolic profile across peel, pulp, and seeds. Molecules 29:3533. https://doi.org/10.3390/molecules29153533
9. Park MK, Cha JY, Kang M-C, Jang HW, Choi Y-S (2024) The effects of different extraction methods on essential oils from orange and tangor: from the peel to the essential oil. Food Sci Nutr 12:804–814. https://doi.org/10.1002/fsn3.3785
10. Pereira Júnior JDA, Costa DS, da Silva ADS, Santos GGD, Santos AFLD, da Silva ADC, do Couto MVS, Cordeiro CAM, Martins ML, Sousa NDC (2025) Enriched diet with orange essential oil Citrus sinensis for tambaqui Colossoma macropomum promotes growth performance and resistance against Aeromonas hydrophila. J Fish Dis 48:e14039. https://doi.org/10.1111/jfd.14039
11. Pimentel LS, Bastos LM, Goulart LR, Ribeiro LNDM (2024) Therapeutic effects of essential oils and their bioactive compounds on prostate cancer treatment. Pharmaceutics 16:583. https://doi.org/10.3390/pharmaceutics16050583
12. Yeasmin MS, Uddin MJ, Dey SS, Barmon J, Ema NT, Rana GMM, Rahman MM, Begum M, Ferdousi L, Ahmed S, Khan MS, Khatun MH, Muzahid AA (2024) Optimization of green microwave-assisted extraction of essential oil from lemon (Citrus limon) leaves: bioactive, antioxidant and antimicrobial potential. Curr Res Green Sustain Chem 8:100413. https://doi.org/10.1016/j.crgsc.2024.100413
13. Pezantes-Orellana C, German Bermúdez F, Matías De la Cruz C, Montalvo JL, Orellana-Manzano A (2024) Essential oils: a systematic review on revolutionizing health, nutrition, and omics for optimal well-being. Front Med (Lausanne) 11:1337785. https://doi.org/10.3389/fmed.2024.1337785
14. Kalompatsios D, Ionescu A-I, Athanasiadis V, Chatzimitakos T, Mantiniotou M, Kotsou K, Bozinou E, Lalas SI (2024) Maximizing bioactive compound extraction from mandarin (Citrus reticulata) peels through green pretreatment techniques. Oxygen 4:307–324. https://doi.org/10.3390/oxygen4030018
15. Mohamed Ahmed IA, Özcan MM, Uslu N, Mohammed BM, Adiamo O (2024) The effects of microwave and oven drying on bioactive compounds individual phenolic constituents and the fatty acid profiles of bitter orange, mandarin and grapefruit peel and oils. Waste Biomass Valor 15:4735–4746. https://doi.org/10.1007/s12649-024-02488-2
16. Anyiam IV, Opara CN, Ezelarry TP (2024) Nutritional properties, antioxidant and antimicrobial activity of seed and peel extracts of grapefruit (Citrus paradisi) on some selected clinical organisms. Asian J Microbiol Biotechnol 9:17–27. https://doi.org/10.56557/ajmab/2024/v9i18564
17. Gilani F, Raftani Amiri Z, Esmaeilzadeh Kenari R, Ghaffari Khaligh N (2023) Investigation of extraction yield, chemical composition, bioactive compounds, antioxidant and antimicrobial characteristics of citron (Citrus medica L.) peel essential oils produced by hydrodistillation and supercritical carbon dioxide. J Food Meas Charact 17:4332–4344. https://doi.org/10.1007/s11694-023-01956-y
18. Ghani A, Taghvaeefard N, Hosseinifarahi M, Dakhlaoui S, Msaada K (2023) Essential oil composition and antioxidant activity of citron fruit (Citrus medica var. macrocarpa Risso.) peel as relation to ripening stages. Int J Environ Health Res 33:1278–1288. https://doi.org/10.1080/09603123.2022.2084514
19. Visakh NU, Pathrose B, Narayanankutty A, Alfarhan A, Ramesh V (2022) Utilization of pomelo (Citrus maxima) peel waste into bioactive essential oils: chemical composition and insecticidal properties. Insects 13:480. https://doi.org/10.3390/insects13050480
20. Li G, Cheng Y, Zhang T, Li Y, Han L, Liang G (2021) Characterization of oxygenated heterocyclic compounds and in vitro antioxidant activity of pomelo essential oil. Drug Des Devel Ther 15:937–947. https://doi.org/10.2147/DDDT.S299678
21. Chakraborty S, Goel K, Rasal V, Paul K, Mandal D (2024) A comprehensive review: exploring bioactive compounds of citrus fruit peels for therapeutic and industrial applications. Food Sci Eng 6:54–69. https://doi.org/10.37256/fse.6120254847
22. Sabry BA, Badr AN, Mohammed DM, Desoukey MA, Farouk A (2024) Validating the protective role of orange and tangerine peel extracts foramending food safety against microorganisms' contamination using molecular docking. Heliyon 10:e27737. https://doi.org/10.1016/j.heliyon.2024.e27737
23. Bozova B, Gölükcü M, Giuffrè AM (2024) The effect of different hydrodistillation times on the

composition and yield of bergamot (*Citrus bergamia* Risso) peel essential oil and a comparison of the cold-pressing method. Flavour Fragr J 39:263–270. https://doi.org/10.1002/ffj.3789

24. Mazzola G, Rondanelli M, Baron G, Zupo R, Castellana F, Clodoveo ML, Gasparri C, Barrile GC, Seniga M, Schiavi LM, Moroni A, Gulec S, Riso P, Perna S (2024) Bergamot (Citrus bergamia), a (poly)phenol-rich source for improving osteosarcopenic obesity: a systematic review. Foods 13:3422. https://doi.org/10.3390/foods13213422
25. Houng P, Ly K, Lay S (2023) Valorization of kaffir lime peel through extraction of essential oil and process optimization for phenolic compounds. J Chem Technol Biotechnol 98:7354. https://doi.org/10.1002/jctb.7354
26. Colecio-Juárez MC, Rubio-Núñez RE, Botello-Álvarez JE (2012) Characterization of volatile compounds in the essential oil of sweet lime. Chil J Agric Res 72:275–280. https://pdfs.semanticscholar.org/6d84/5f590e26e595e22992d8908256f9e626843d.pdf. Accessed 10 Jan 2025
27. de Almeida BS, de Carvalho Gomes QC, de Menezes RDF et al (2025) Innovative products of mayonnaise-type and dessert creams developed with the kefir-based beverage of Colocasia esculenta. J Culin Sci Technol:1–19. https://doi.org/10.1080/15428052.2025.2497349
28. Ningrum A, Suna G, Karaçil-Ermumcu MŞ, Yeşilyurt N, Yılmaz B (2024) Citrus seed waste and circular bioeconomy: insights on nutritional profile, health benefits, and application as food ingredient. In: Valorization of fruit seed waste from food processing industry. Elsevier, pp 67–84. https://doi.org/10.1016/b978-0-443-15535-2.00005-0
29. Dhakal M, Ghimire S, Karki G, Deokar GS, Al-Asmari F, Nirmal NP (2024) Fruit waste: potential bio-resource for extraction of nutraceuticals and bioactive compounds. In: Methods and protocols in food science. Springer US, New York, pp 299–332. https://doi.org/10.1007/978-1-0716-3601-5_13
30. Thong A, Basri N, Chew W (2024) Comparison of untargeted gas chromatography-mass spectrometry analysis algorithms with implications to the interpretation and putative identification of volatile aroma compositions. J Chromatogr A 1713:464519
31. Arena A, Zoccali M, Bonaccorsi I, Mondello M, Tranchida PQ, Mondello L (2024) Determination of mineral oil hydrocarbon contamination in Citrus essential oils by using on-line liquid-gas chromatography: critical aspects. Anal Bioanal Chem 416:801–808. https://doi.org/10.1007/s00216-023-04873-7
32. Khanem A, Karim N, Ullah I, Younas F (2025) Essential oils of Citrus limon, Cymbopogon citratus, and Lavandula officinalis disrupt E. coli biofilms by inducing cellular damage. Biol Futur 76:343–358. https://doi.org/10.1007/s42977-025-00255-5
33. Dharmamoorthy G (2025) A comprehensive review on hyphenated techniques in pharmaceutical analysis. Asian J Pharm 19:6052
34. Amala Dev AR, Joseph SM (2025) Chemical diversity and correlation analysis of volatiles from Citrus peels: evaluation of toxicity, antioxidant and microbial inhibition efficacy. Microbe 7:100370
35. Michiu D, Socaciu M-I, Fogarasi M, Jimborean AM, Ranga F, Mureşan V, Semeniuc CA (2022) Implementation of an analytical method for spectrophotometric evaluation of total phenolic content in essential oils. Molecules 27:1345. https://doi.org/10.3390/molecules27041345
36. Azizian T, Alirezalu A, Hassani A, Bahadori S, Sonboli A (2021) Phytochemical analysis of selected Nepeta species by HPLC-ESI-MS/MS and GC–MS methods and exploring their antioxidant and antifungal potentials. J Food Meas Charact 15:2417–2429. https://doi.org/10.1007/s11694-021-00819-8
37. Hermes L, Römermann J, Cramer B, Esselen M (2021) Quantitative analysis of β-asarone derivatives in Acorus calamus and herbal food products by HPLC-MS/MS. J Agric Food Chem 69:776–782. https://doi.org/10.1021/acs.jafc.0c05513
38. Liu X, Zhang Q, Zong C, Gai H (2025) Digital immunoassay for proteins: theory, methodology, and clinical applications. Anal Chem 97:9077–9110. https://doi.org/10.1021/acs.analchem.4c05421
39. Kato-Noguchi H, Kato M (2025) Pesticidal activity of citrus fruits for the development of sustainable fruit-processing waste management and agricultural production. Plants 14:754. https://doi.org/10.3390/plants14050754
40. Ahmed IAM, Özcan MM, Uslu N, Al-Juhaimi F, Wei R (2025) The role of water- and ultrasononic bath processes on bioactive properties, phenolic profiles, fatty acids, lipid indices of roasted and unroasted bitter orange seeds. Waste Biomass Valor 16:4321–4332
41. Aćimović M, Jeremić JS, Cvetković M, Lončar B, Pezo L (2025) Comparative analysis of essential oils and hydrolates from aniseed, coriander, and fennel fruits. Chem Biodivers 22:e202500022

42. Spedale C, Frazzini S, Rossi L, Dell'Anno M, Chiofalo V (2025) Functional characterization of antimicrobial and antioxidant properties of whole lemon pulp and depectinized lemon pulp as circular by-products in animal nutrition. J Agric Food Res 21:101998. https://doi.org/10.1016/j.jafr.2025.101998
43. Khaled R, Mohamed S, Mohamed A, Khairy A, Elhariry H, Bakry A, Elsenosy NK, Ebeed N, Salem SS, ElNoby T, Abu-Hussien SH (2025) Optimization, characterization and biosafety of oregano, rosemary and mint oil mixture against Penicillium digitatum in citrus using L-optimal mixture design. AMB Express 15:14. https://doi.org/10.1186/s13568-024-01806-0
44. Kaur S, Panesar PS, Chopra HK, Singh V (2025) Exploring the bioactive potential of Citrus reticulata pomace: extraction, characterization, and applications. Food Hum 4:100518
45. Yeddes W, Bettaieb Rebey I, Yazidi R, Ben Abdennebi A, Hammami M, Ben Farhat M, Saidani Tounsi M (2025) Eco-friendly extraction of hesperidin from citrus peels: a comparative study of Soxhlet, ultrasound-assisted, and microwave-assisted methods for improved yield and antioxidant properties. Int J Environ Health Res:1–12. https://doi.org/10.1080/09603123.2025.2485426
46. Sharma R (2024) Extraction methods for essential oils from citrus peels. In: The functional foods. Apple Academic Press, New York, pp 273–294. https://doi.org/10.1201/9781032684055-16
47. Cucinotta L, Cafeo G, Alibrando F, Russo M, Sciarrone D, Mondello L, Mondello M (2025) Characterization of oxygen heterocyclic compounds and volatile fraction of Citrus australasica finger lime peel essential oil exploiting a multi-technique chromatographic approach. J Essent Oil Res 37:46–55
48. Liu S, Li S, Ho C-T (2025) Oxygen heterocyclic compounds (OHCs) in lemon oils are important markers in targeted HPLC analyses for authentication. Food Sci Human Wellness 14:9250027. https://doi.org/10.26599/fshw.2024.9250027
49. Pu J, Cui J, Yang H, Cao J, Xiao S, Cheng G (2025) The anti-inflammatory effects and molecular mechanism of Citri Reticulatae Pericarpium essential oil: a combined GC-MS and network pharmacology study. Foods 14:1455
50. El Hachlafi N, Elbouzidi A, Batbat A, Taibi M, Jeddi M, Addi M, Naceiri Mrabti H, Fikri-Benbrahim K (2024) Chemical composition and assessment of the anti-inflammatory, antioxidant, cytotoxic and skin enzyme inhibitory activities of Citrus sinensis (L.) Osbeck essential oil and its major compound limonene. Pharmaceuticals (Basel) 17:1652. https://doi.org/10.3390/ph17121652
51. Żukowska G, Durczyńska Z (2024) Properties and applications of essential oils: a review. J Ecol Eng 25:333
52. Kaya O, Bozkurt A, Karakus S, Daler S, Yilmaz T, Turan M (2024) Essential oils in post-harvest disease management: metabolic impact on Narince (Vitis vinifera L. cv) grapes against Botrytis cinerea. Physiol Mol Plant Pathol 132:102318. https://doi.org/10.1016/j.pmpp.2024.102318
53. Hübschmann HJ (2025) Handbook of GC-MS: fundamentals and applications. John Wiley & Sons
54. Sandhya H, Sah PJ (2025) Conventional techniques for extracting essential oil: Steam distillation and hydrodistillation, in: Methods and Protocols in Food Science, Springer US, New York, NY. 21–41. https://doi.org/10.1007/978-1-0716-4634-2_2
55. Huo Y, Deng W, Sun X, Zhou L, Zhang Q, Hu J (2024) Extract toolkit for essential oils: state of the art, trends, and challenges. Food Chem 461:140854
56. Purnima TN, Sharma AK, Singh A, Kumar V, Tomer PK, Babele (2026) Phytochemicals Extraction Quantification Isolation and Characterization Chromatography (HPLC-DAD) Methods for Analyzing Alkaloids from Plants Springer US New York NY 293-299
57. Wang H-S, Li N, Zhang C-D, Ling Y-D, Yan F (2024) Chromatographic analysis of paeonol in natural plants: an undergraduate laboratory experiment for comparative observation of HPLC and GC performance. J Chem Educ 101:4425–4432. https://doi.org/10.1021/acs.jchemed.4c00820
58. Description of chromatographic performance in reversed-phase liquid chromatography based on a generalised model. Part II: Prediction of serial-column coupling
59. Sangeetha KPP, Srivastav S, Sadanandan PP, Srivastav B, Srivastava S, Karunanithi (2025) Essential Oil Extraction from Food By-Products Application of Essential Oil in Food Systems Springer US New York NY 303-314
60. Zhang Z-Y, Liu Y-F, Zhang S-J, Zhang P-P, Shen X-X, Lan J-L, Mao Z-J, Zhang M-J, Ruan Y-P, Zhang X (2025) Rescue of CUMS-induced HPA axis hyperfunction and hypothalamic synaptic deficits by Citrus aurantium L. cv. Daidai essential oil via the cAMP/PKA/Grin2b pathway. J Ethnopharmacol 343:119423. https://doi.org/10.1016/j.jep.2025.119423

INDEX

Shweta Suri et al. (eds.), *Citrus Fruit Processing By-Products: Technologies and Basic Protocols*, Methods and Protocols in Food Science, https://doi.org/10.1007/978-1-0716-5088-2,

MIX
Papier aus verantwortungsvollen Quellen
Paper from responsible sources
FSC® C105338

If you have any concerns about our products,
you can contact us on
ProductSafety@springernature.com

In case Publisher is established outside the EU,
the EU authorized representative is:
Springer Nature Customer Service Center GmbH
Europaplatz 3, 69115 Heidelberg, Germany

Printed by Libri Plureos GmbH
in Hamburg, Germany